Multiplication Puzzles for Kids
30 Puzzles using facts 0-12
for kids ages 7-10

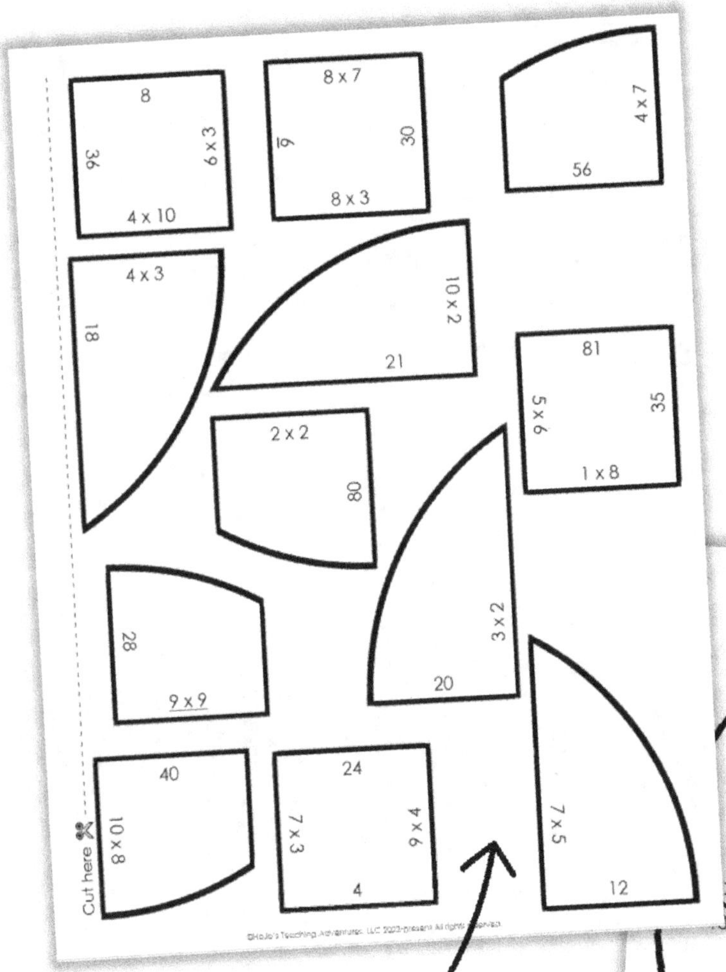

The goal is simple - practice multiplication facts by cutting out the puzzle pieces and putting them back together by answering the multiplication problems. Once complete, you'll reveal a fun image!

Go from this...

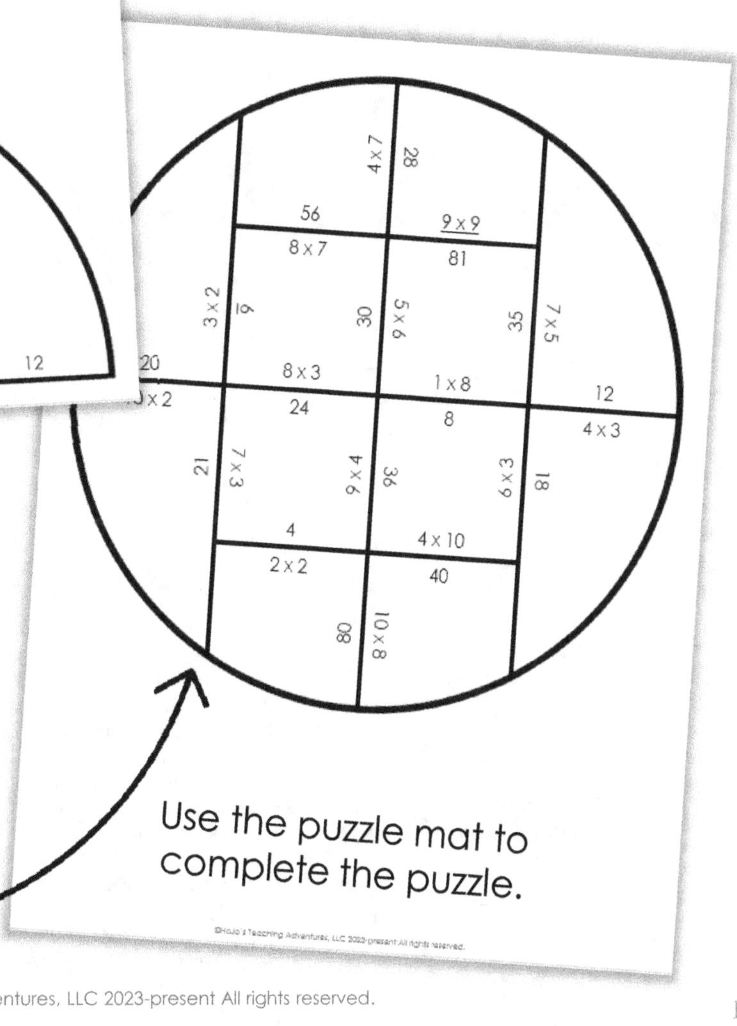

...to this!

Use the puzzle mat to complete the puzzle.

Table of Contents

Table of Contents

Instructions for Use

These puzzles were created to keep the practice of multiplication facts **hands-on**, **engaging**, and **FUN**. Help kids work on and memorize their multiplication facts 0-12. Choose the puzzle you want, cut on the dotted line, and hand it to the child. The child will cut out the individual puzzle pieces and then put the puzzle back together.

If you want a bit more support for the child, also give them the puzzle mat so they can see the shape they are working on. These puzzle mats are optional, but they do offer more support for kids who are new to these puzzles or those who need a bit more help as they work through the multiplication facts.

There are **30 different puzzle shapes** to choose from, so you're sure to find something that will work for ANY time of year!

Here are the puzzle options included: snowman, mitten, penguin, mug, groundhog, heart, clover, horseshoe, bunny, egg, bee, raindrop, flip flop, apple, leaf, acorn, pumpkin, candy corn, cat, turkey, football, sneaker, gingerbread man, candy cane, stocking, tooth, circle, square, triangle, and star.

Ideas for Extra Support:
- Use the included multiplication charts for students who may be struggling. You can choose to give them the chart that includes numbers 1-12, or give them just the 11/12 chart so they need to know numbers 1-10 on their own.
- Allow the use of a calculator. (The memorization will come, but some students need support for longer than others.)
- Make sure they use the included puzzle mat for each shape!

Extended Learning:
- Save the puzzles and let students complete them more than one time.
- Have students draw their own shape, make cross lines to create puzzles like the ones in this book, and then write out their own multiplication problems. Cut and you'll have another puzzle to put together.

 Answer keys for all 30 puzzles have been included at the end of this book.

I hope you enjoy these puzzles as much as I have enjoyed creating them! This teacher turned mom THANKS YOU for your interest in my work! Happy learning!!

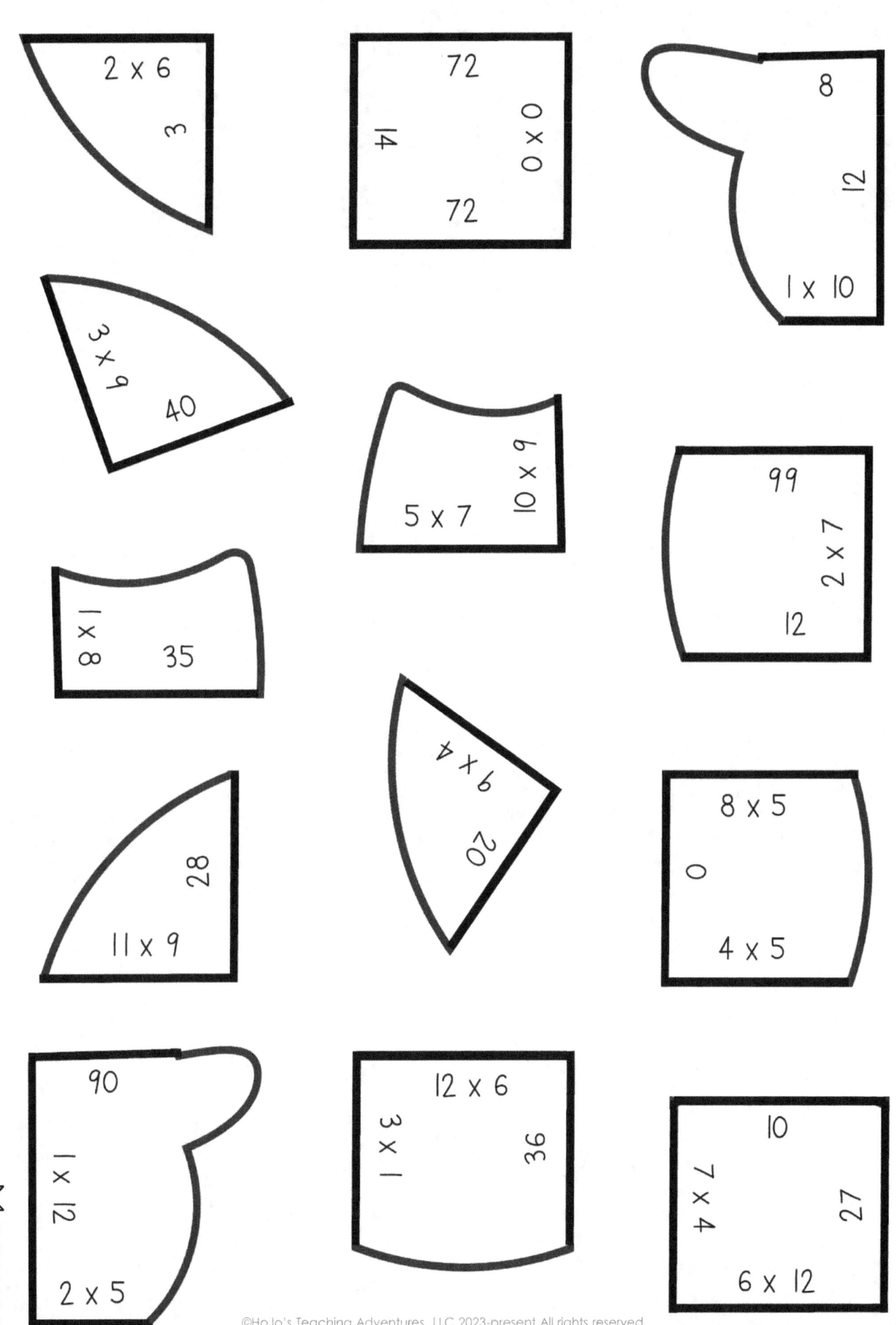

2 x 6

3

72

14

0 x 0

72

8

12

1 x 10

3 x 9

40

10 x 6

5 x 7

99

2 x 7

12

1 x 8

35

28

11 x 9

9 x 4

20

8 x 5

0

4 x 5

90

1 x 12

2 x 5

12 x 6

3 x 1

36

10

7 x 4

27

6 x 12

Use the puzzle mat to complete the puzzle.

64
11 x 3

0
2 x 2

70
7 x 11
48

1 x 8
9

8 x 8
0 x 0
28

4 x 2
18
24
5 x 7

44
12 x 11
9 x 11
77

3 x 3
132
2 x 6

33
9 x 3
1 x 12

12
25
3 x 6

35
7 x 10
9 x 9
99

8 x 9
12

5 x 5
70
10 x 3

7 x 4
27
48
8

4
12 x 4
72

30
11 x 4
8

8 x 9
6 x 4
22
1 x 0

2 x 11
9 x 0
54
10 x 7

0

Use the puzzle mat to complete the puzzle.

1

44

55   22

7 x 7

72   108

4 x 11

9 x 6

8

5 x 11

36

12 x 9

36

24

4 x 9

36

3 x 12

2 x 11

11 x 6

6 x 9

8 x 8

1 x 8

50

14

66

1 x 1

5 x 10

3 x 8

54   5 x 2

2 x 7   6 x 9

49

10

Use the puzzle mat to complete the puzzle.

9 x 7
6 x 9
64
4 x 4

2 x 1
10 x 5
5 x 10
12 x 5
88

36
3 x 3
60
63

1 x 11
3 x 9
2 x 10

20
5 x 4
80
6 x 6

0 x 0
27
7 x 9

0
1 x 0

6 x 9
11
16

54
0

4 x 4
20
48

50

8 x 6
9
7 x 7

63
10 x 8
2

49
8 x 8

11 x 8
24

Use the puzzle mat to complete the puzzle.

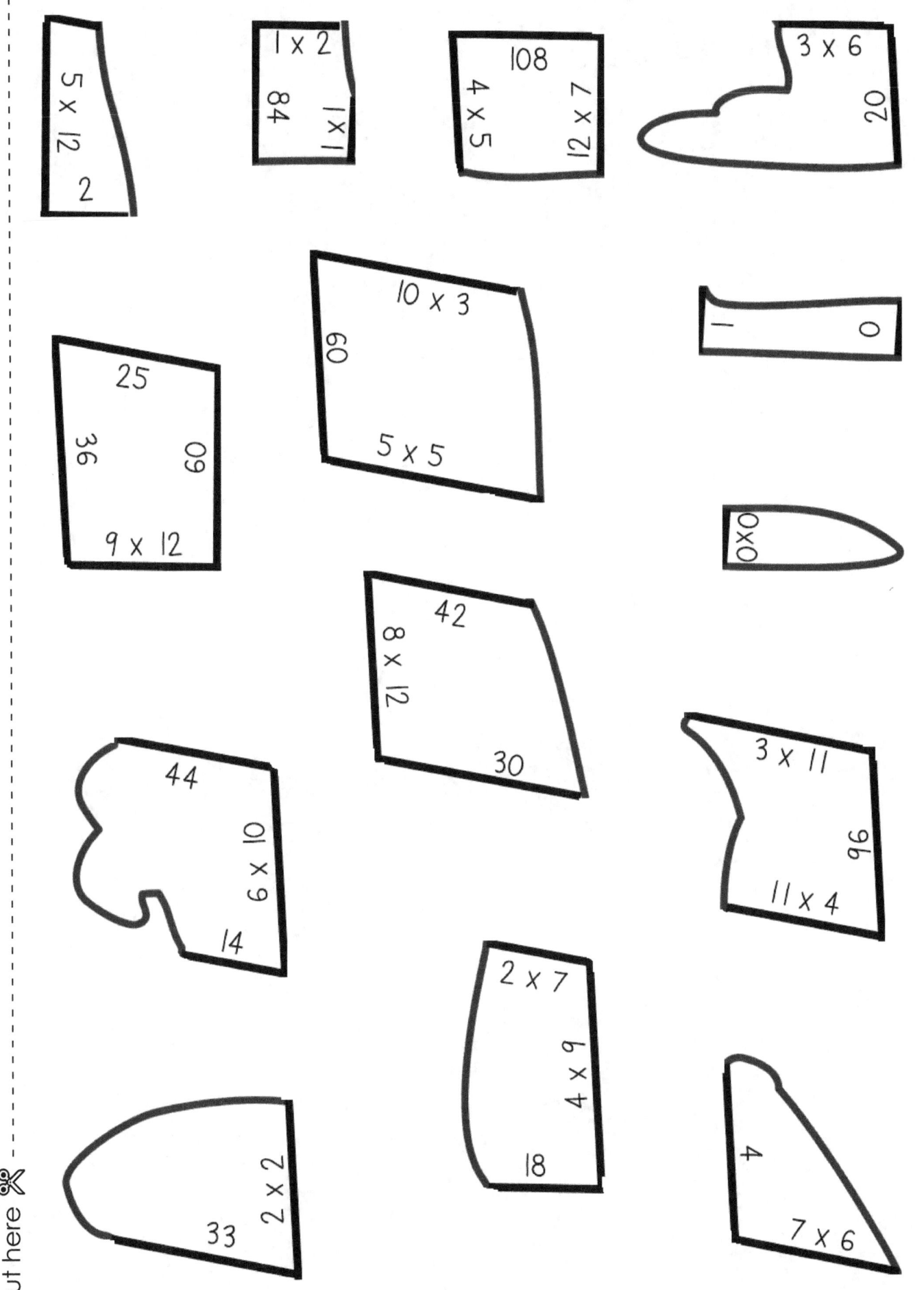

5 x 12

2

1 x 2

84

1 x 1

108

7 x 2

4 x 5

12 x 9

3 x 6

20

25

36

09

9 x 12

10 x 3

60

5 x 5

1

0

0 x 0

42

8 x 12

30

44

01

6 x 9

14

3 x 11

96

11 x 4

2 x 7

4 x 9

18

2 x 2

33

2 x 7

4

7 x 6

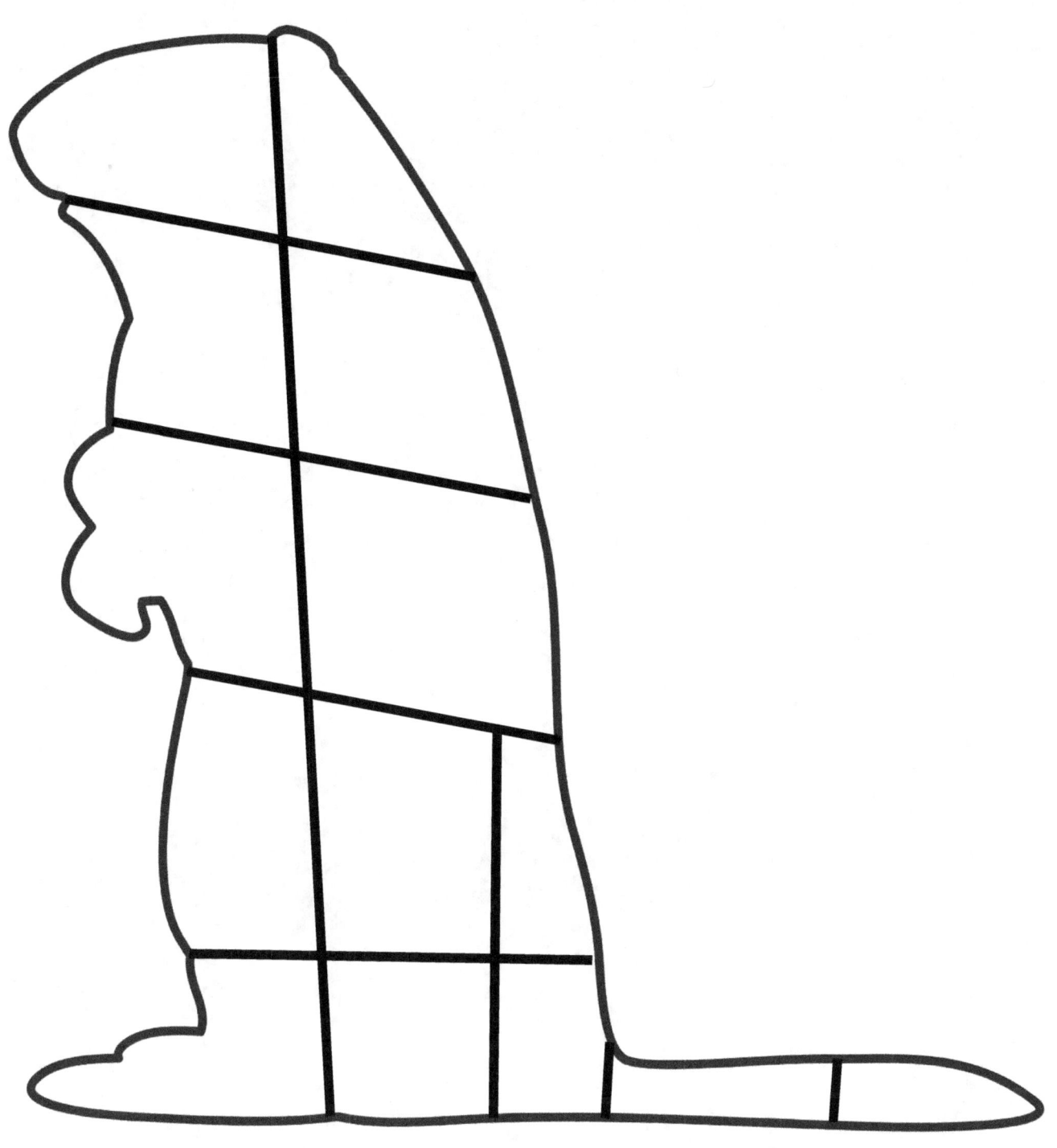

Use the puzzle mat to complete the puzzle.

7 x 8

5

99

72

6 x 4

3 x 5

12 x 6

30

56

0

1 x 5

64

24

28

8 x 4

10 x 3

12 x 0

6 x 11

4 x 7

45

15

36

5 x 9

11 x 12

8 x 8

9 x 9

3

12

16

63

90

3 x 1

9 x 2

9 x 7

10 x 9

132

4 x 4

32

Use the puzzle mat to complete the puzzle.

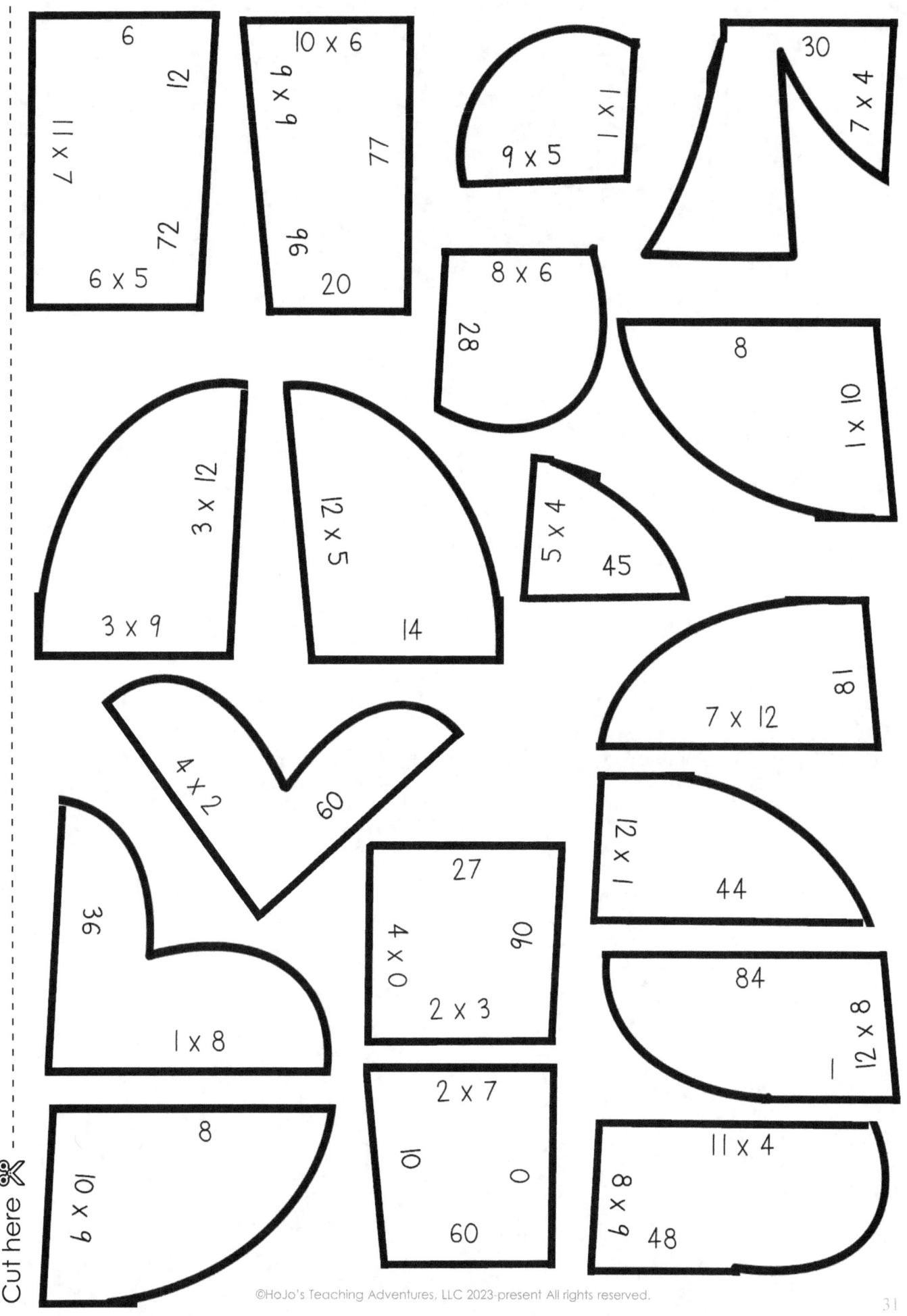

Cut here ✂

Use the puzzle mat to complete the puzzle.

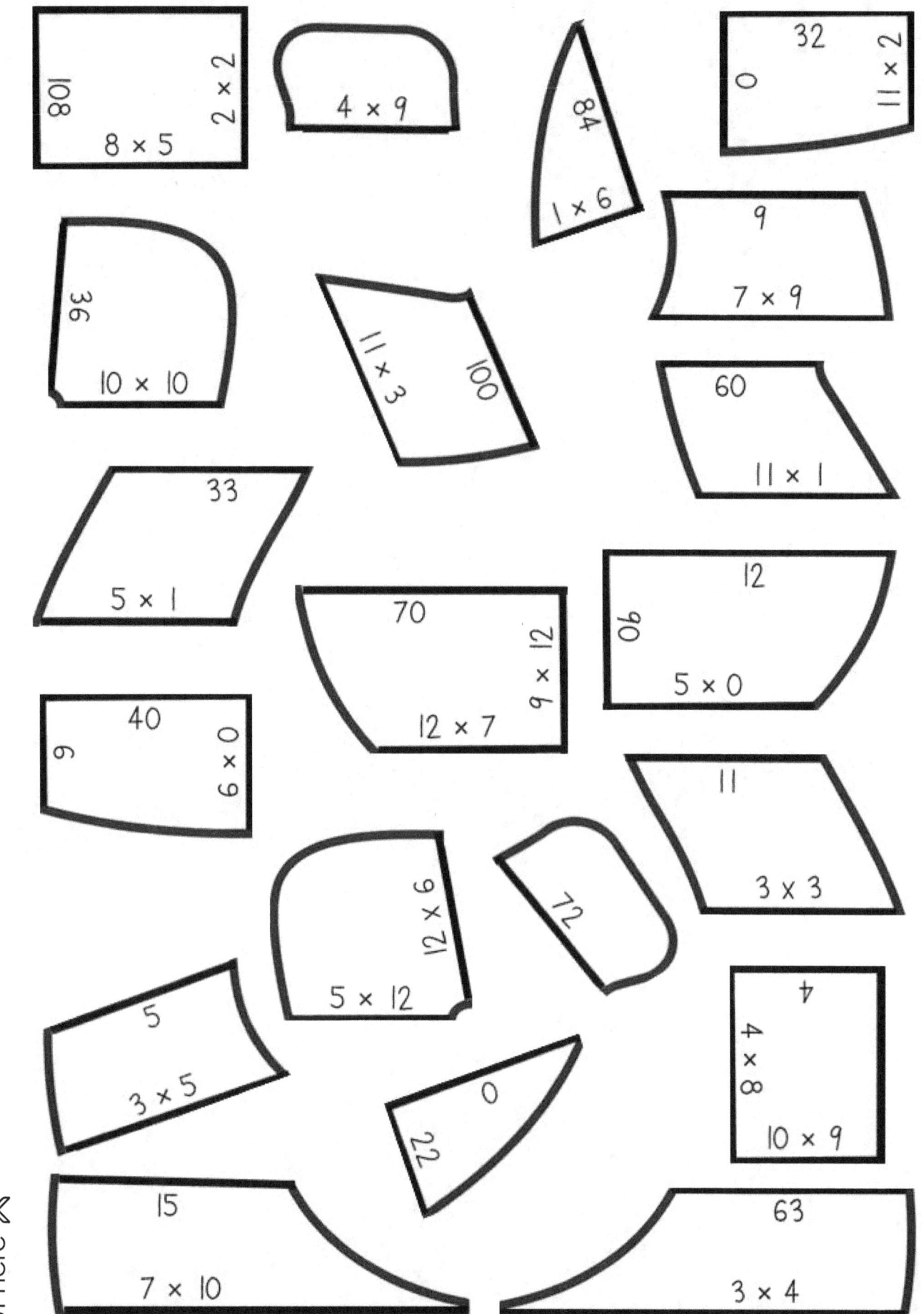

108
2 × 2
8 × 5

4 × 9

84
1 × 6

32
0
11 × 2

36
10 × 10

11 × 3
100

9
7 × 9

60
11 × 1

33
5 × 1

70
12 × 9
12 × 7

12
90
5 × 0

40
6
0 × 9

11
3 × 3

12 × 9
5 × 12

72

0
22

5
3 × 5

4
4 × 8
10 × 9

15
7 × 10

63
3 × 4

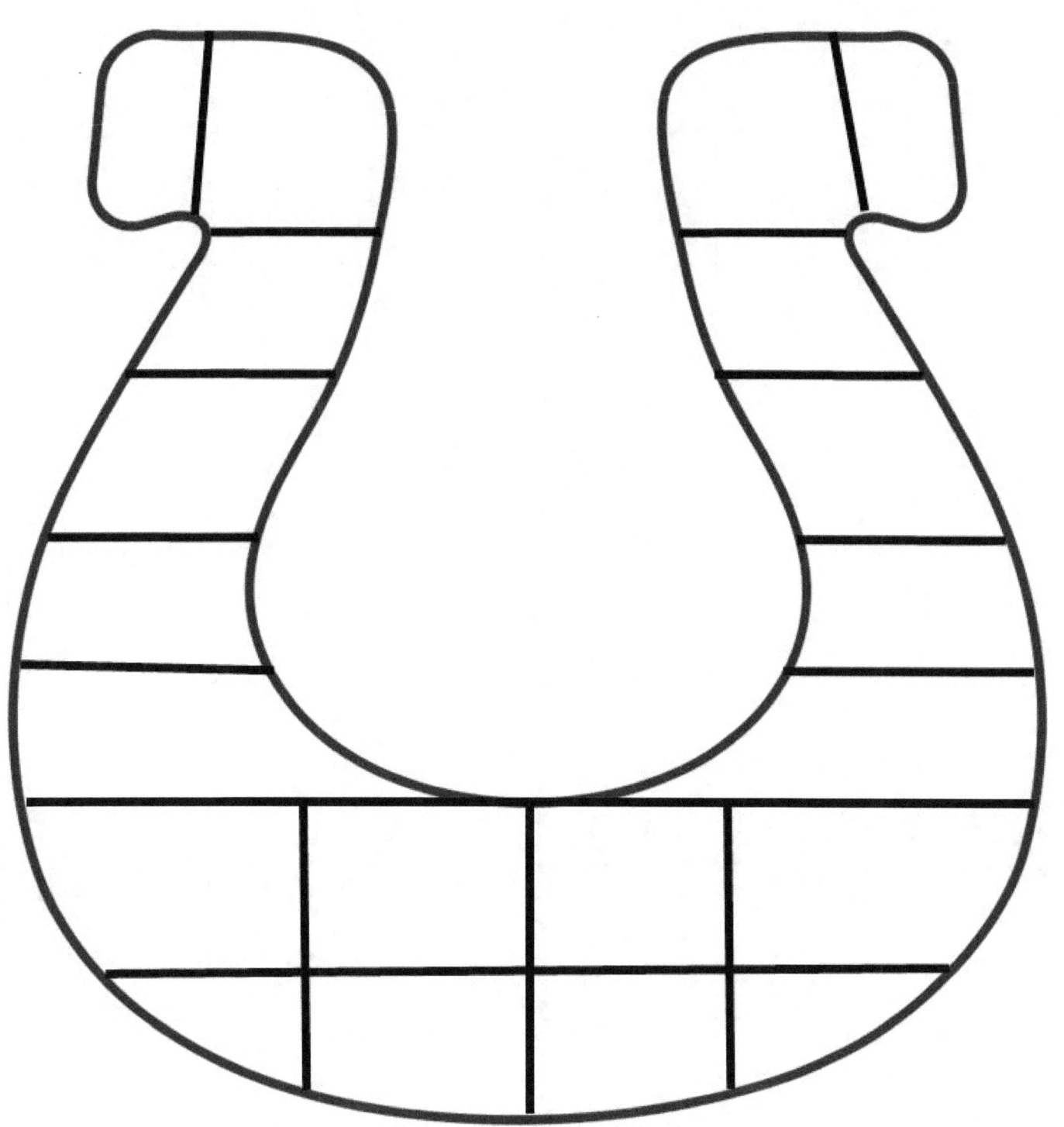

Use the puzzle mat to
complete the puzzle.

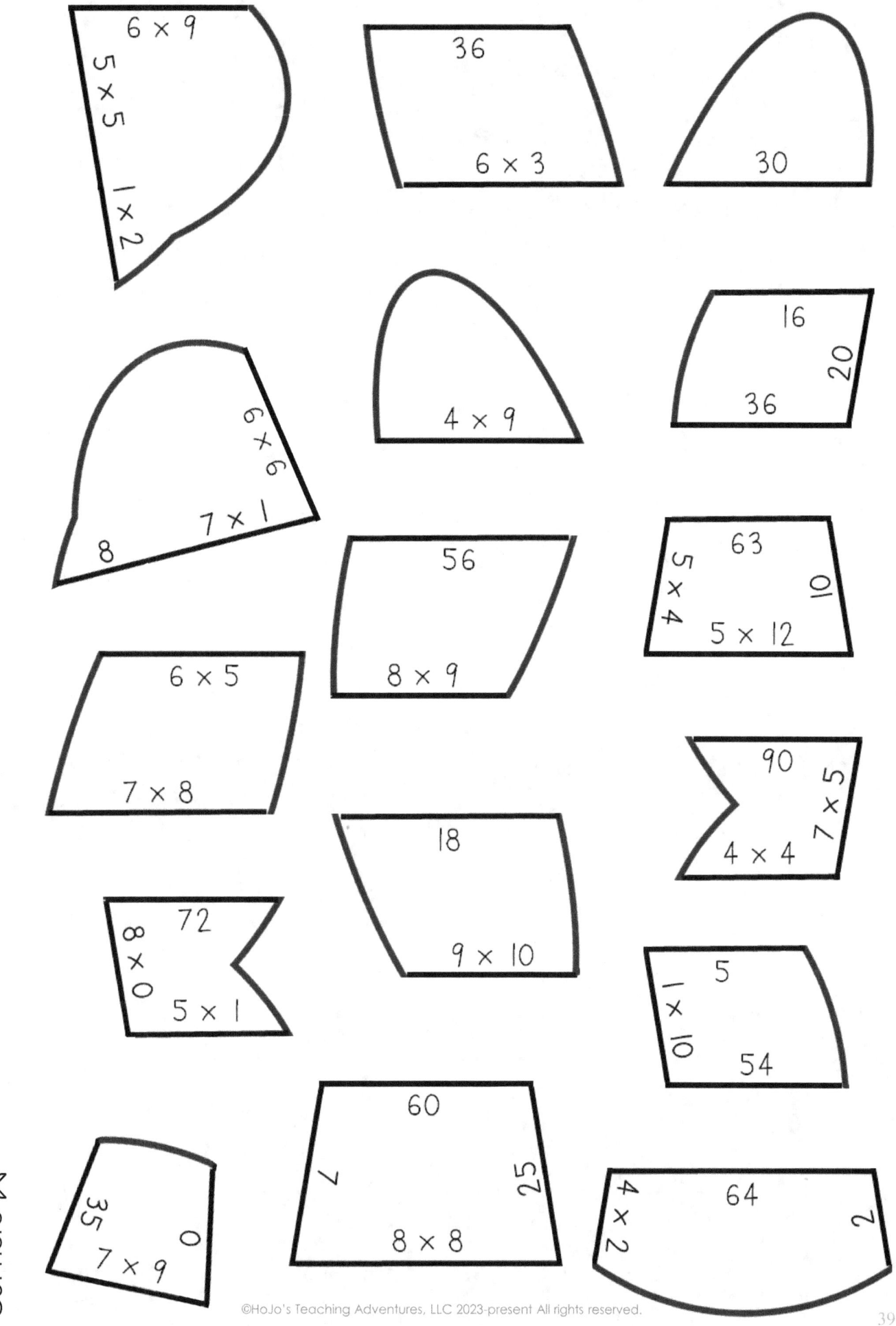

6 × 9
5 × 5
1 × 2

36
6 × 3

30

4 × 9

16
20
36

6 × 6
7 × 1
8

56
8 × 9

63
5 × 4
10
5 × 12

6 × 5
7 × 8

18
9 × 10

90
7 × 5
4 × 4

72
8 × 0
5 × 1

5
1 × 10
54

35
7 × 9
0

60
7
25
8 × 8

64
4 × 2
2

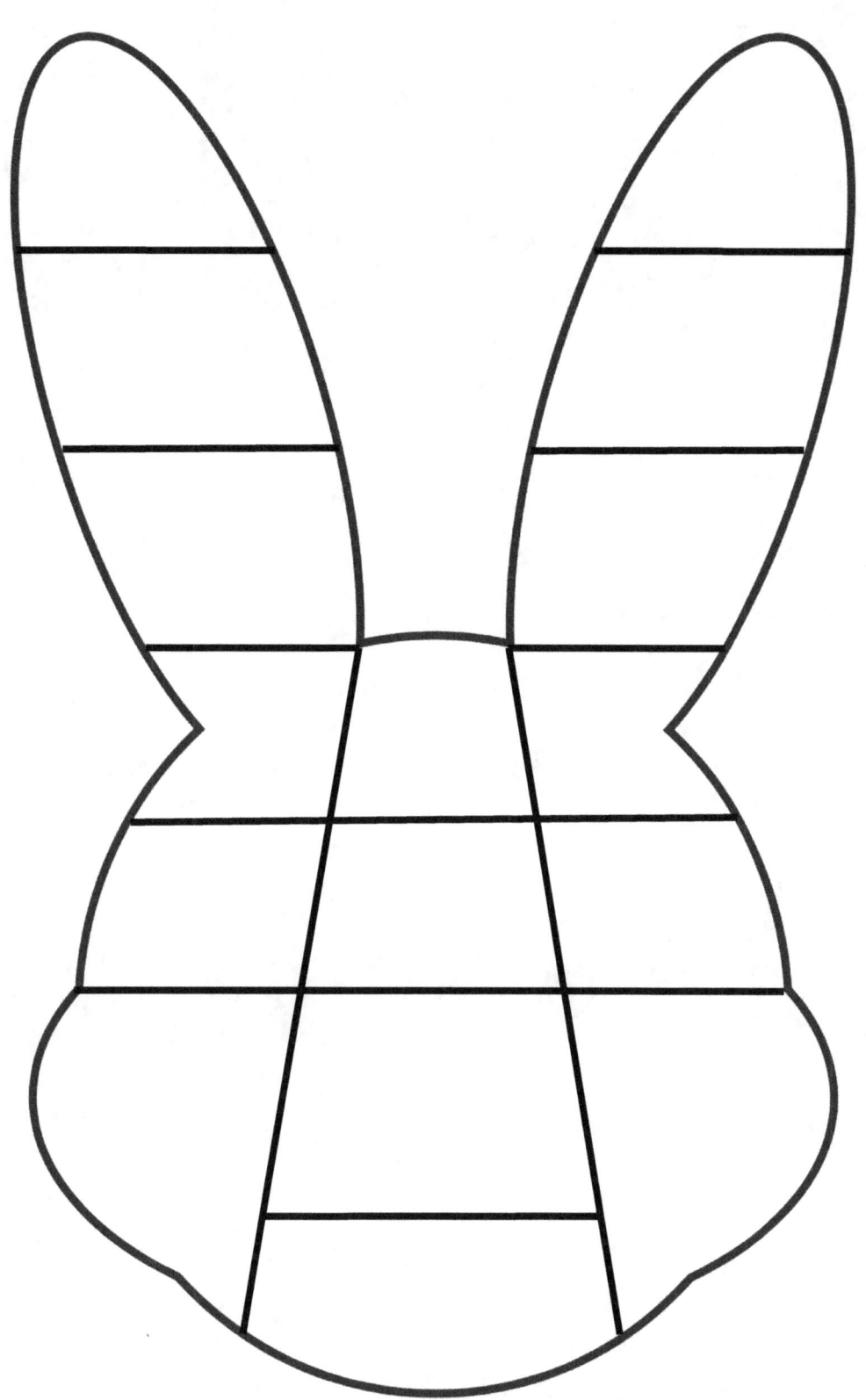

Use the puzzle mat to complete the puzzle.

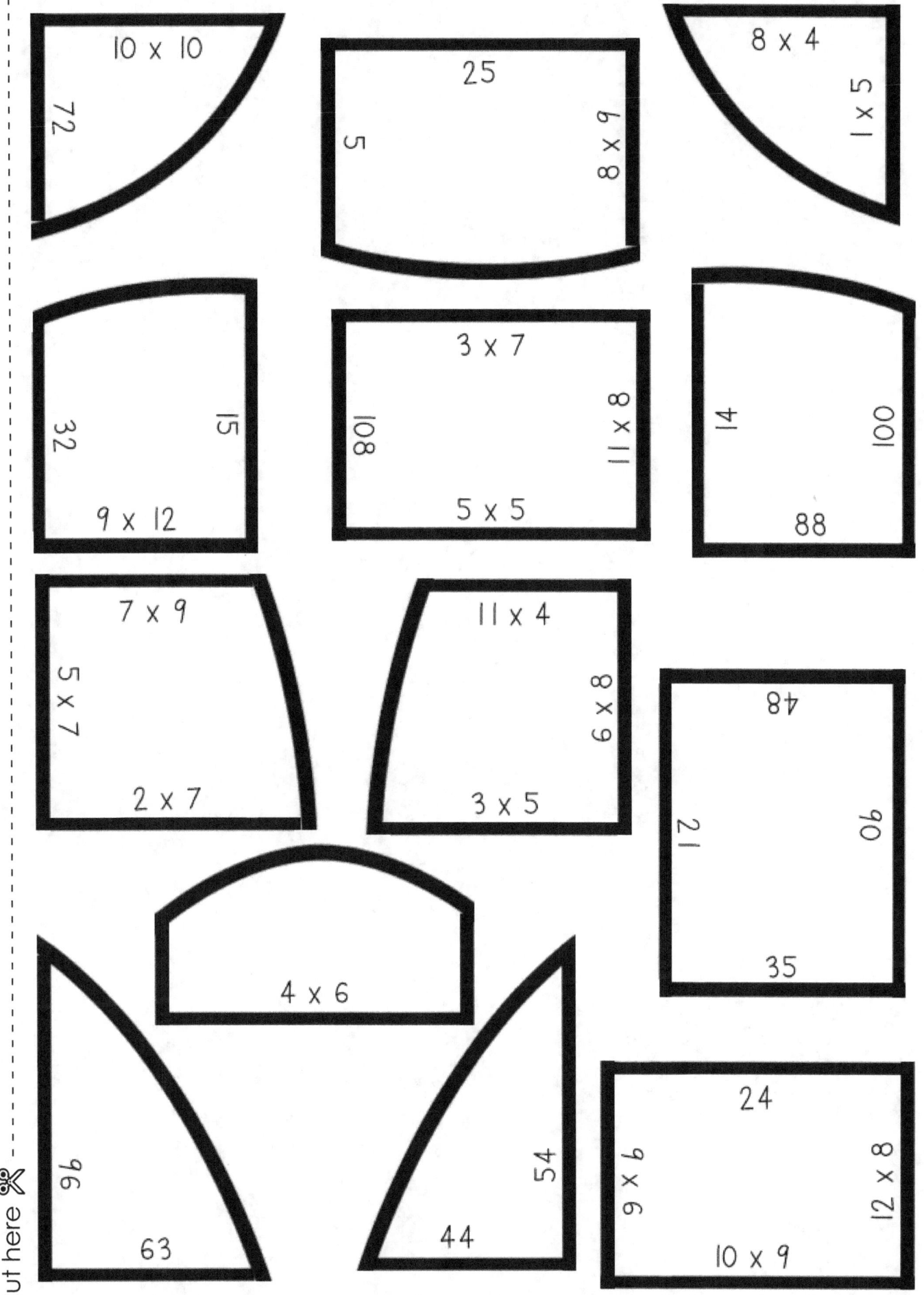

Use the puzzle mat to complete the puzzle.

9 x 0   15

14   2 x 2

9 x 9   27

10   4 x 6   72   11 x 3

33   7 x 6   0

7 x 2   110

8 x 6   28   8 x 2

10 x 4   3 x 9   42

48   3

36   11 x 10   8 x 12

4 x 7   2 x 6   24   40

4   6 x 6

30   16   1 x 10

12 x 6   5 x 3

3 x 1   5 x 6

96   12   81

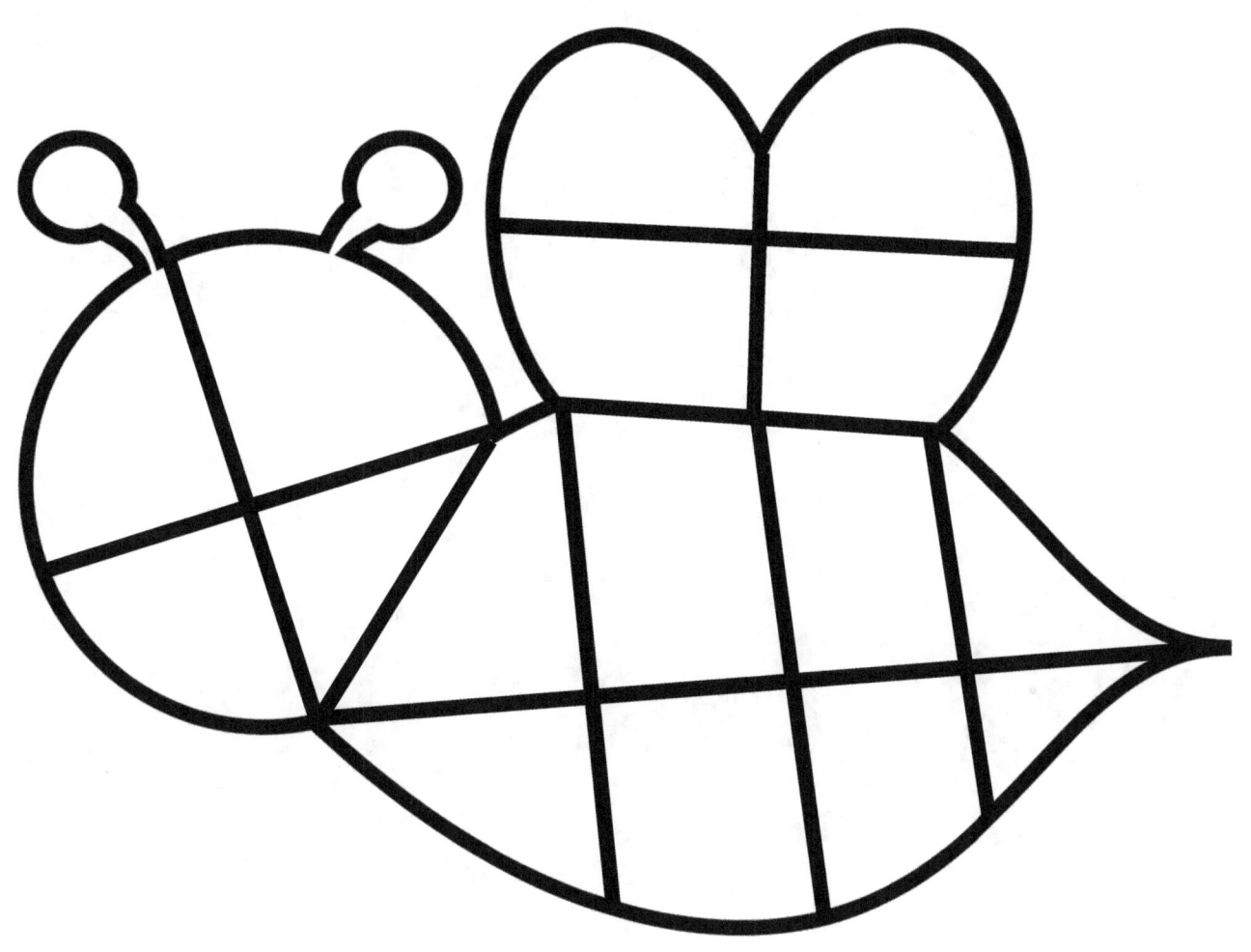

Use the puzzle mat to complete the puzzle.

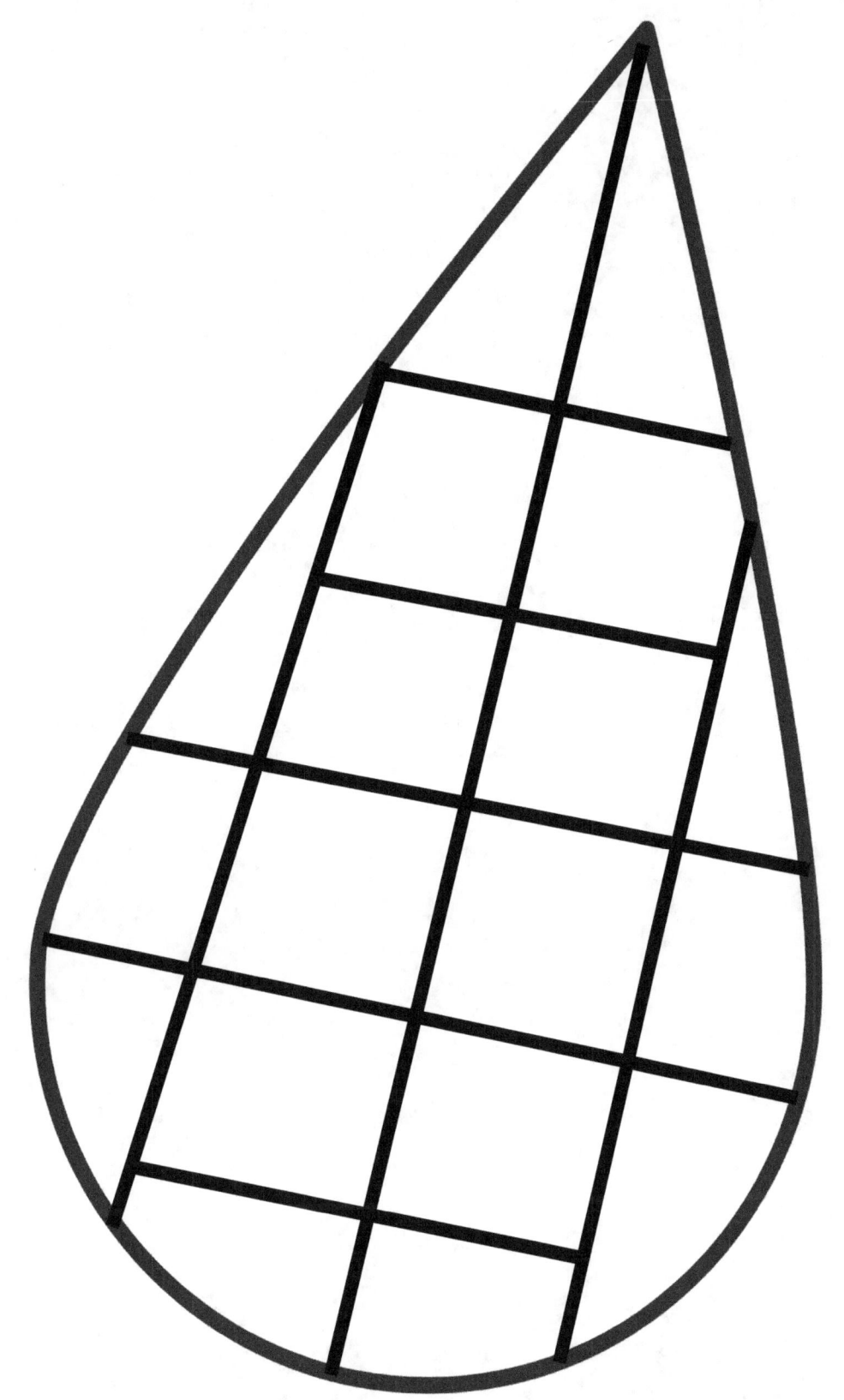

Use the puzzle mat to complete the puzzle.

54   0   3 x 8   14

6 x 6
11 x 4

0
44

45
2 x 7

48
7 x 8
12 x 0

9 x 9
4

1 x 4
56
36

0 x 0
18
4 x 12

8 x 4
10   24

12 x 12
32

10 x 7
9 x 9

144
5 x 9

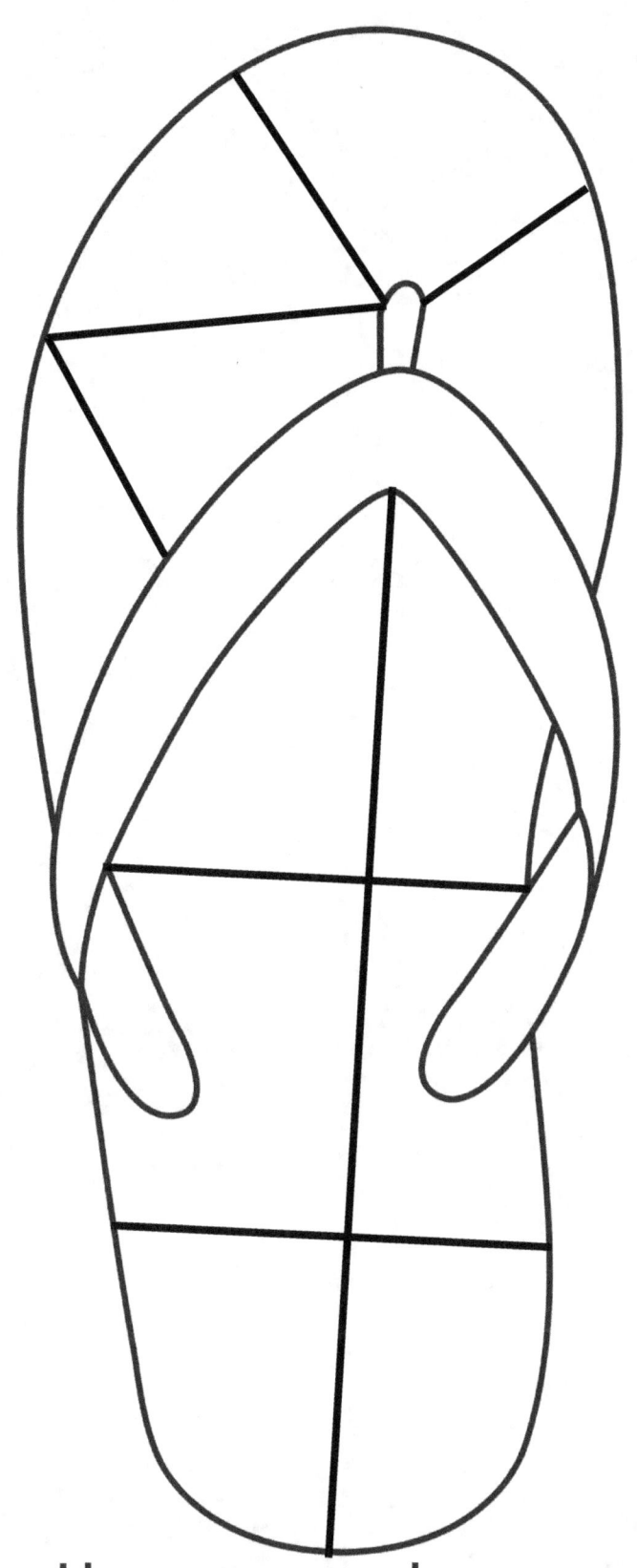

Use the puzzle mat to complete the puzzle.

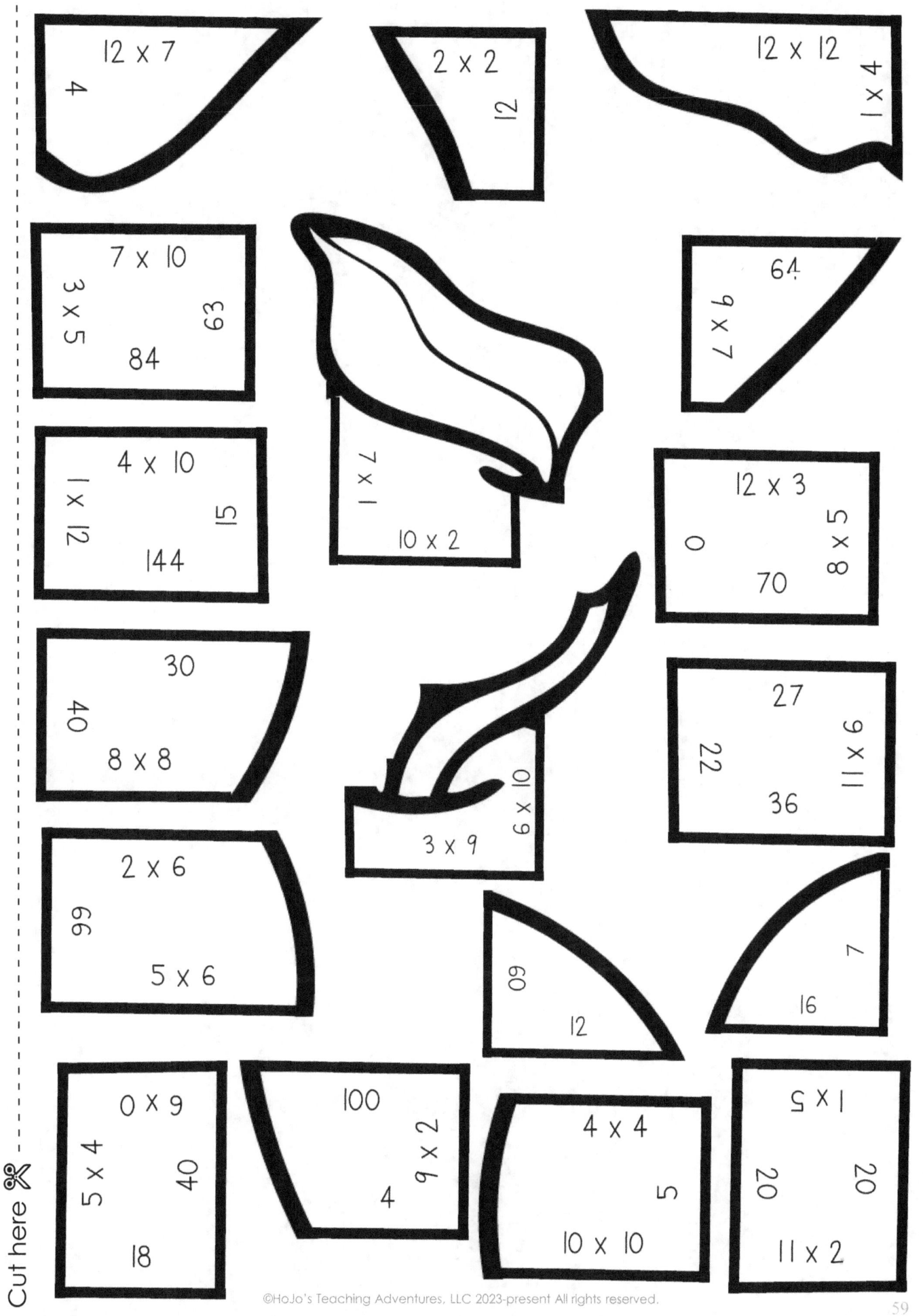

12 x 7
4

2 x 2
12

12 x 12
1 x 1

7 x 10
3 x 5
63
84

64
9 x 7

4 x 10
1 x 12
15
144

7 x 1
10 x 2

12 x 3
0
5 x 8
70

30
40
8 x 8

27
22
9 x 11
36

2 x 6
66
5 x 6

3 x 9
6 x 10

60
12

7
16

9 x 0
5 x 4
40
5 x 5
18

100
9 x 2
4

4 x 4
5
10 x 10

1 x 5
20
20
20
11 x 2

Use the puzzle mat to complete the puzzle.

4 x 2
54

56
24

8 x 7
72

2 x 6
25
144

10 x 12
9 x 6
5 x 9

36
9 x 12
45

5 x 5
3 x 4

9 x 8
40
9 x 9

7 x 2
108

30
99
8

81
4 x 4
55

3 x 8
10 x 4
12
10

21
9 x 11
11 x 11
48
120

11 x 12
8 x 9
5 x 12
12 x 3

12 x 12
3 x 0
8 x 8
7 x 3

11 x 5
60
14

12
64
5 x 6

5 x 2
16
0
132

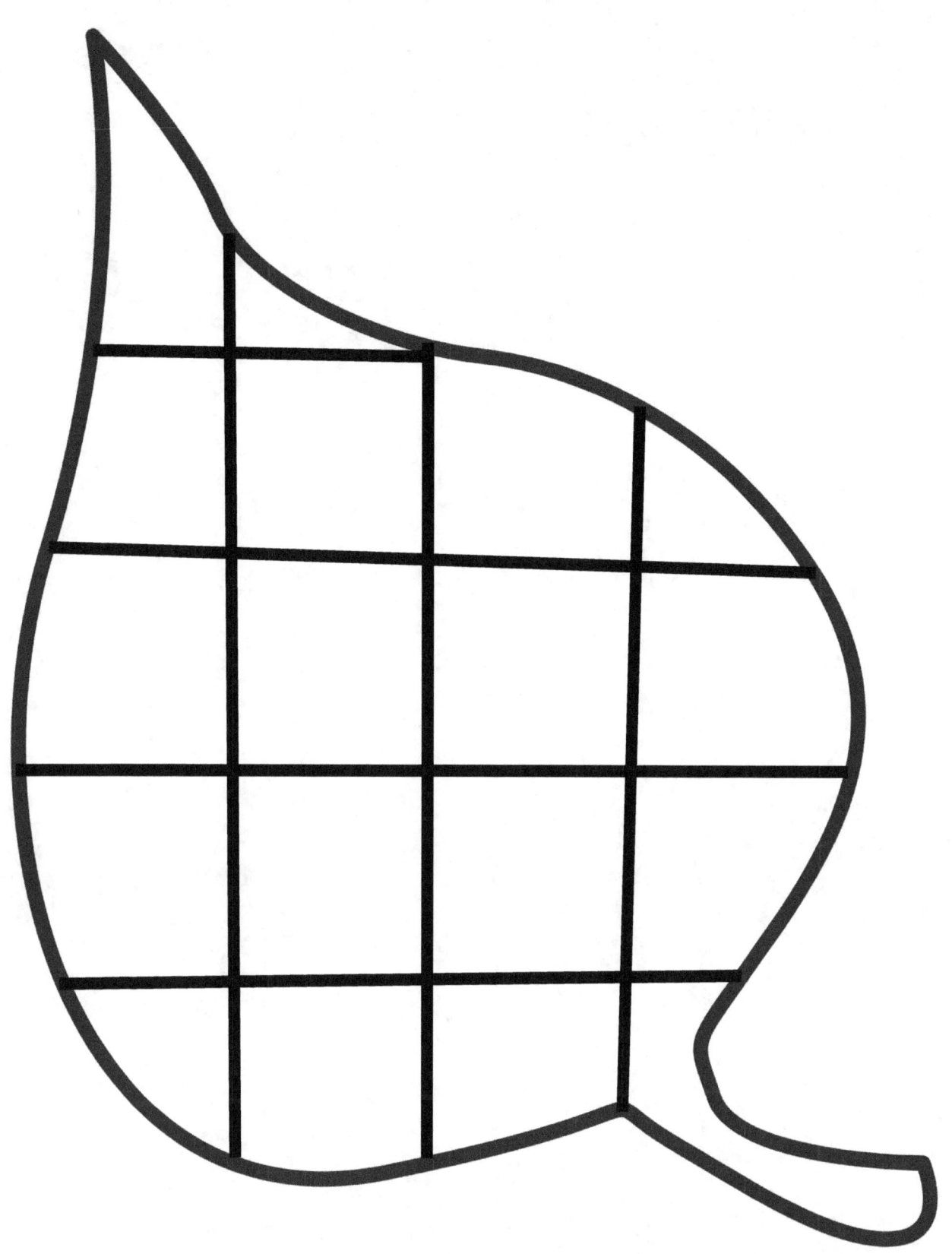

Use the puzzle mat to complete the puzzle.

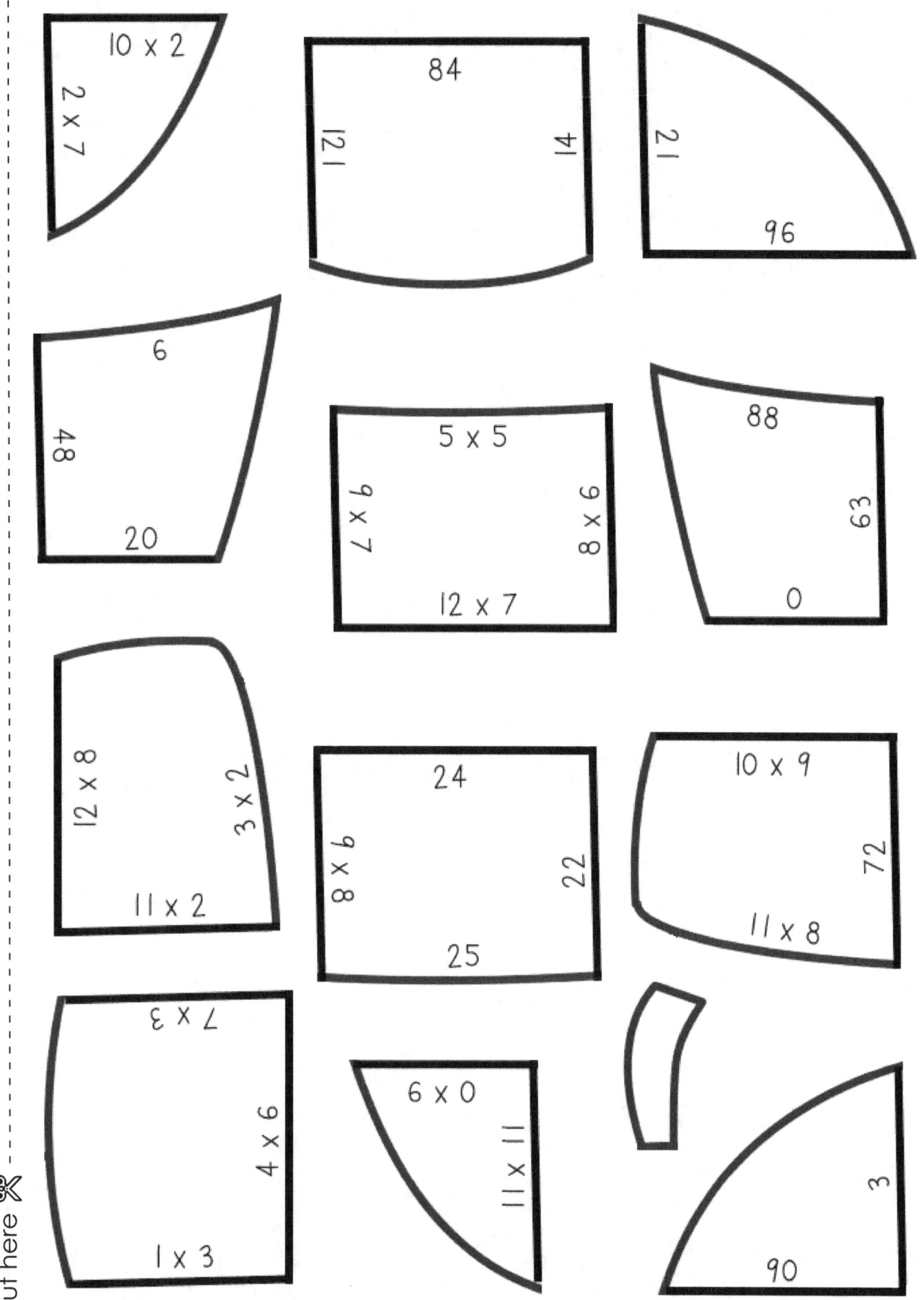

10 x 2
2 x 7

84
121
14

21
96

6
48
20

5 x 5
9 x 7
8 x 6
12 x 7

88
63
0

8 x 12
3 x 2
11 x 2

24
9 x 8
22
25

10 x 9
72
11 x 8

7 x 3
4 x 6
1 x 3

6 x 0
11 x 11

3
90

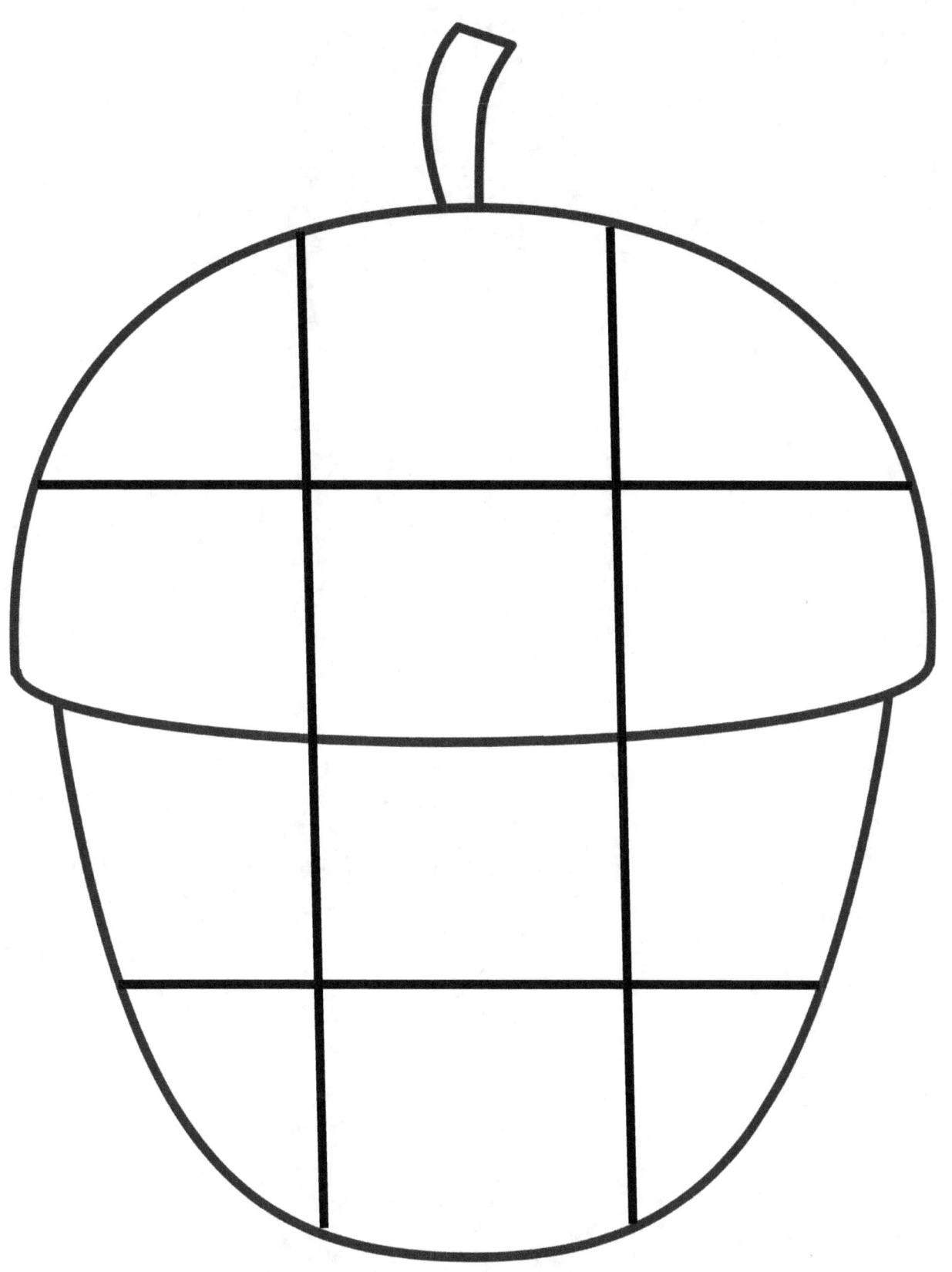

Use the puzzle mat to complete the puzzle.

63
15

32
144
3 x 5

1 x 8
96
12 x 12

72
12 x 8

7 x 10
8 x 4
42
0

5 x 9
5 x 5
70
8

4
12 x 6
25

48
99

11 x 11
9 x 7
10 x 0

27
12 x 4
6 x 7

14
9 x 3
45

7 x 2
2 x 2

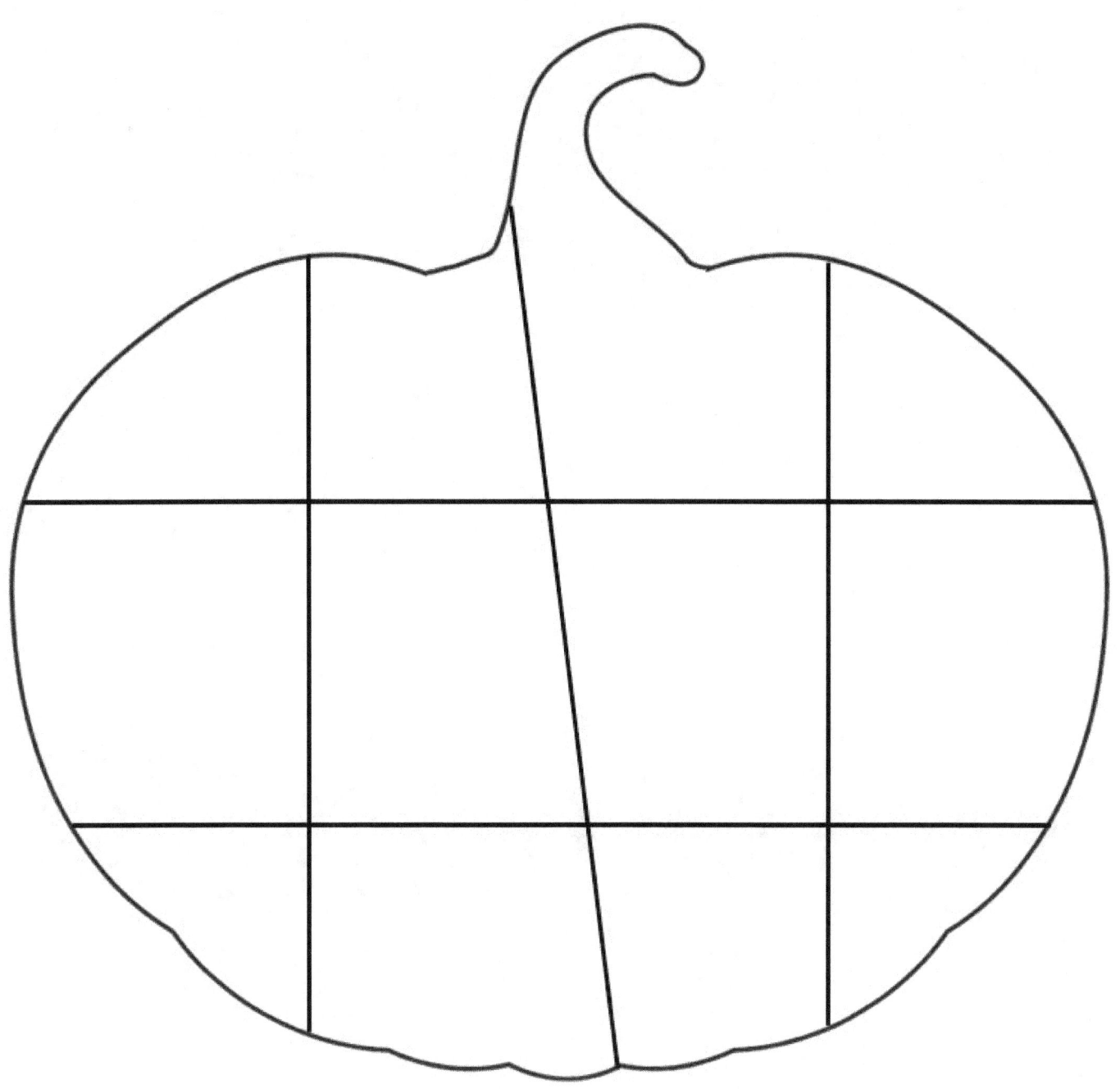

Use the puzzle mat to complete the puzzle.

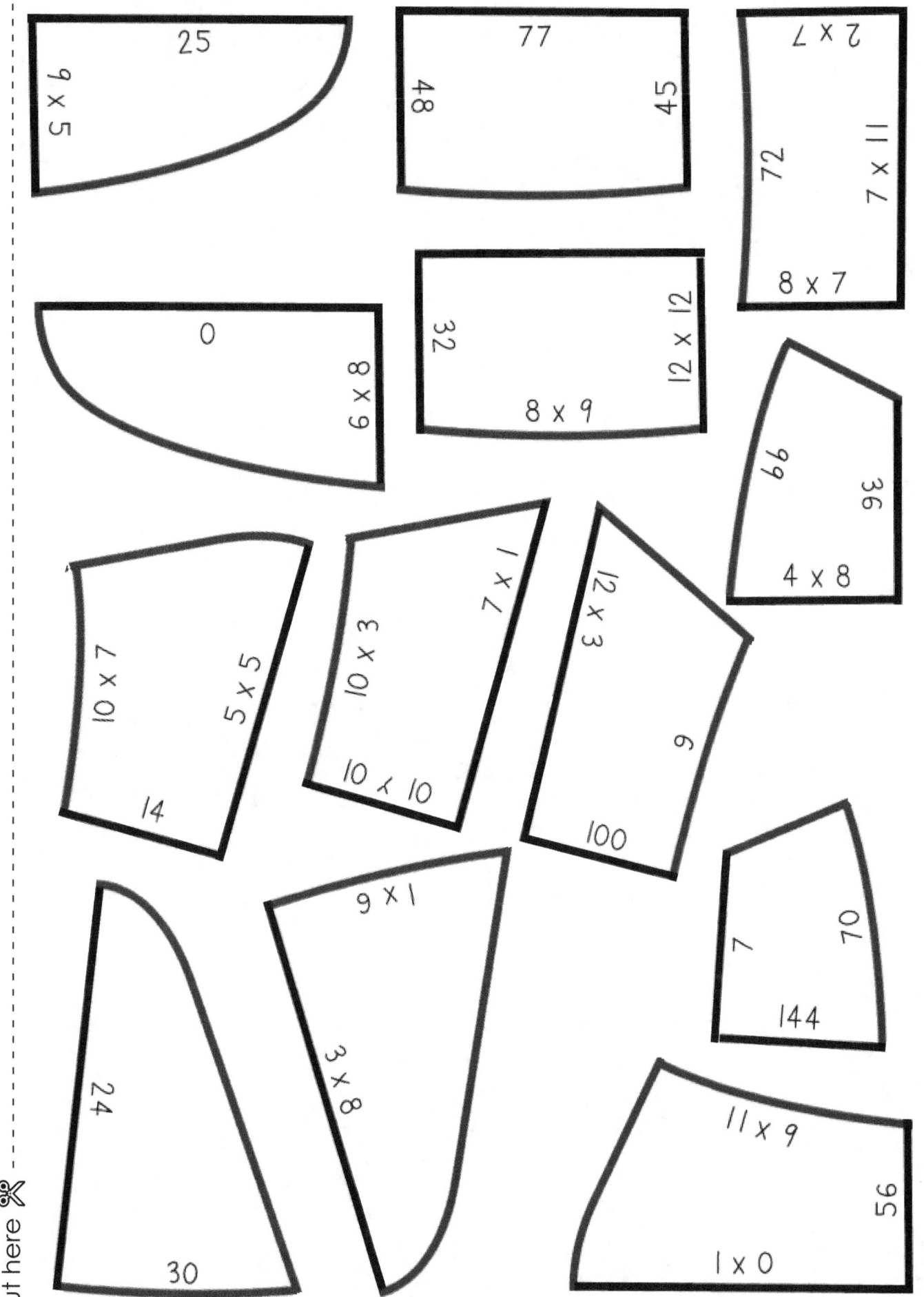

25

9 x 5

77

48

45

2 x 7

7 x 11

72

8 x 7

0

8 x 6

32

12 x 12

8 x 9

36

66

8 x 4

7 x 10

5 x 5

14

7 x 1

10 x 3

10 x 10

12 x 3

6

100

70

7

144

9 x 1

3 x 8

24

30

11 x 9

95

0 x 1

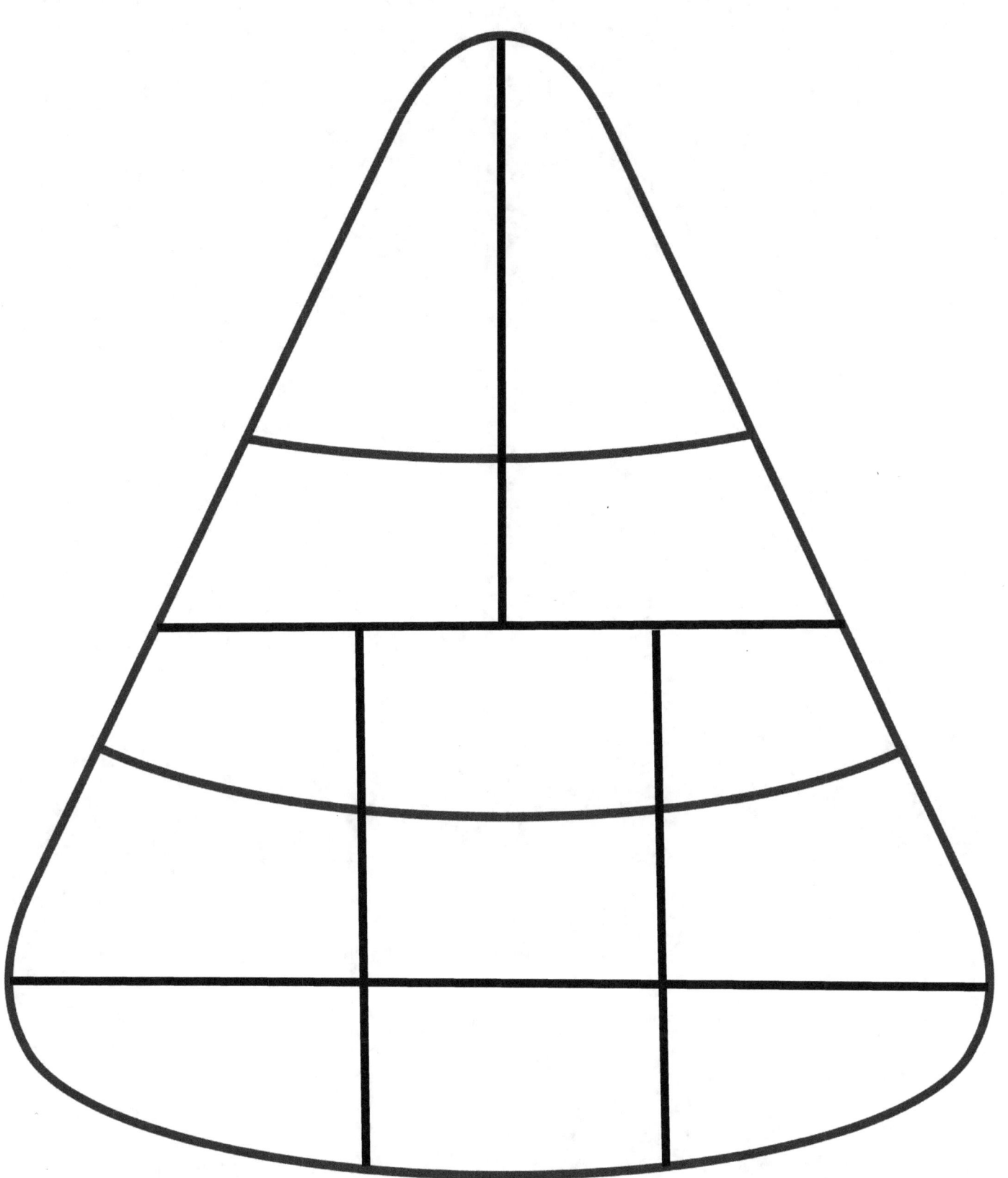

Use the puzzle mat to
complete the puzzle.

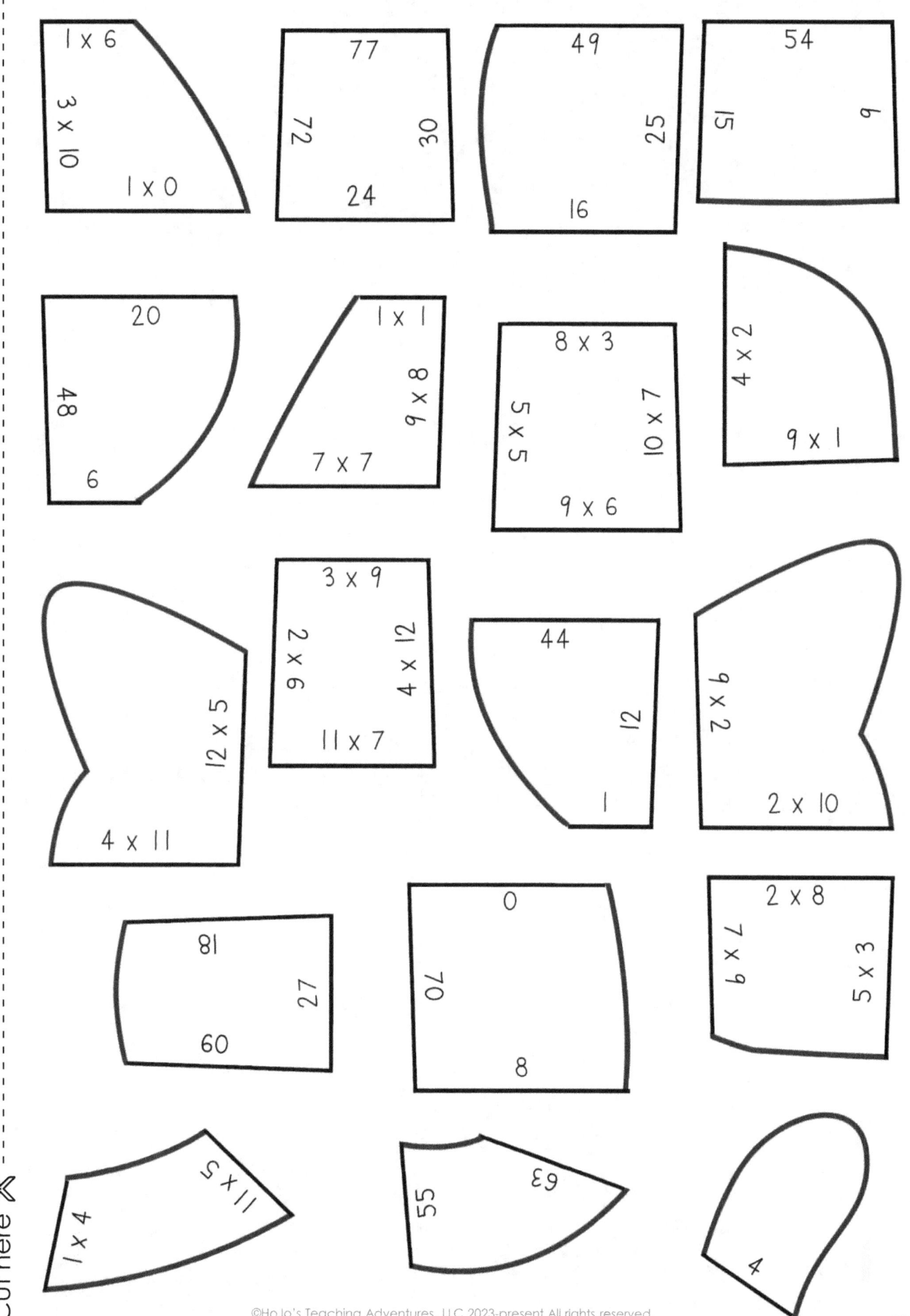

1 x 6
3 x 10
1 x 0

77
72
30
24

49
25
16

54
15
9

20
48
6

1 x 1
9 x 8
7 x 7

8 x 3
5 x 5
10 x 7
9 x 6

4 x 2
9 x 1

3 x 9
2 x 6
4 x 12
11 x 7

44
12
1

9 x 2
2 x 10

12 x 5
4 x 11

81
27
60

0
70
8

2 x 8
7 x 9
5 x 3

1 x 4
11 x 5

55
63

4

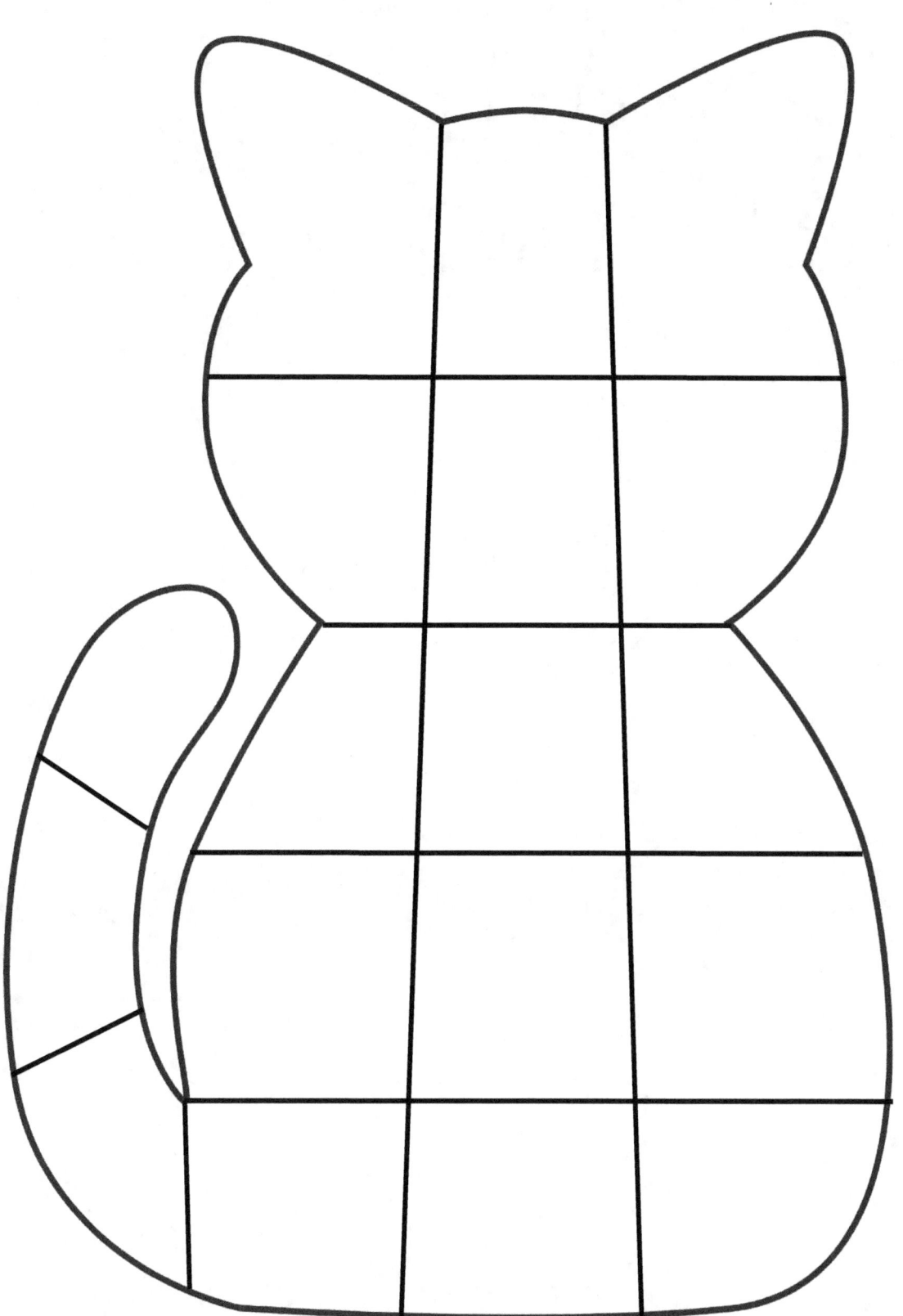

Use the puzzle mat to complete the puzzle.

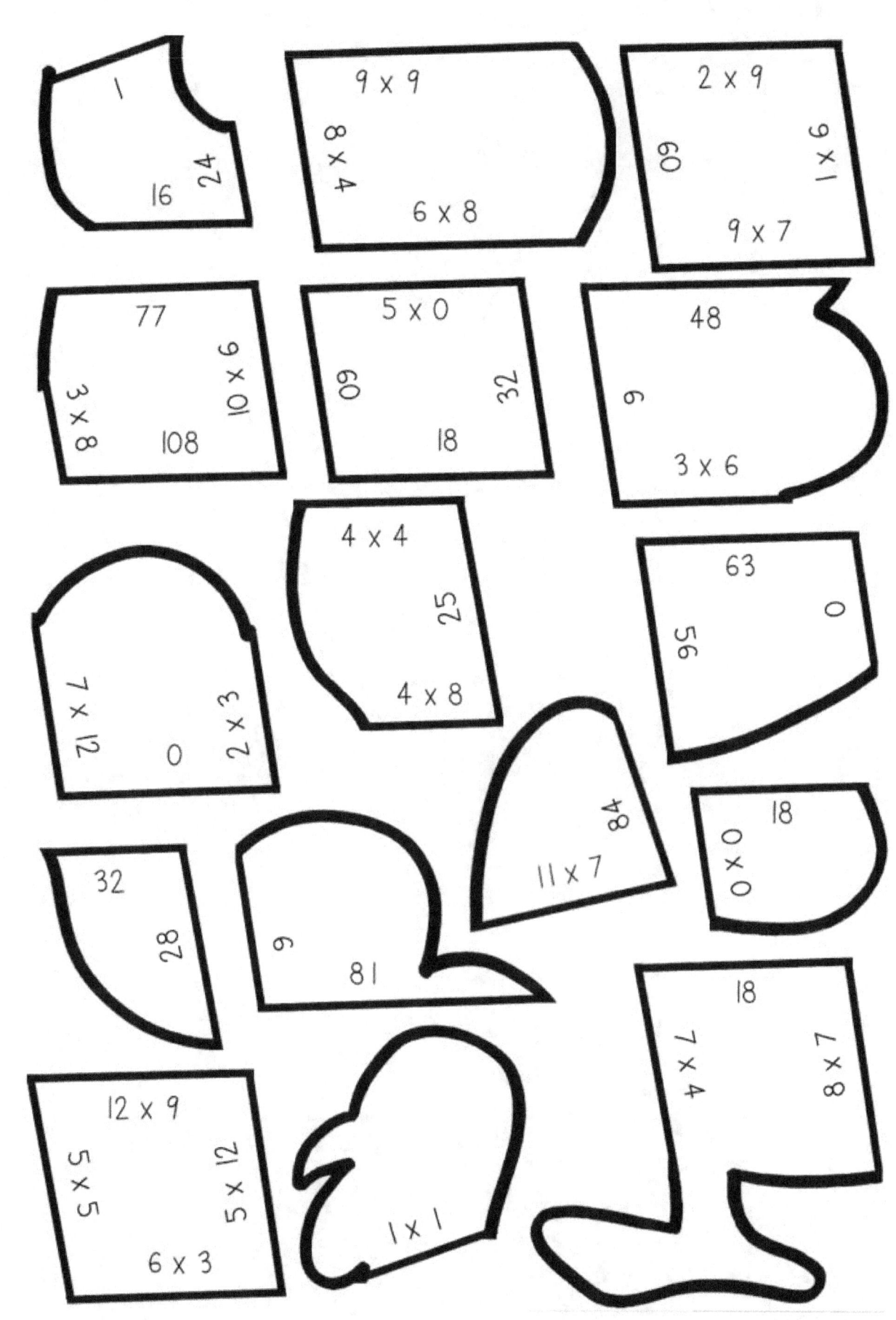

Use the puzzle mat to complete the puzzle.

42   3 x 12

5 x 7   72   8 x 6   10

2 x 8   6 x 6   10 x 3   49

4 x 3   16

132   44   3 x 4   25

5 x 5   3   6 x 10

8 x 3   30   99   11 x 12

7 x 7   12 x 9   11 x 4   9 x 7

10 x 9   28   36   35

70   4 x 5   3 x 8

63   1 x 3

8   60   55

24   90

5 x 11   48

20   9 x 11   4 x 7   22

12   7 x 10   24

2 x 11   12   12 x 6   1 x 8

180   0

6 x 7   2 x 5

36   1 x 0

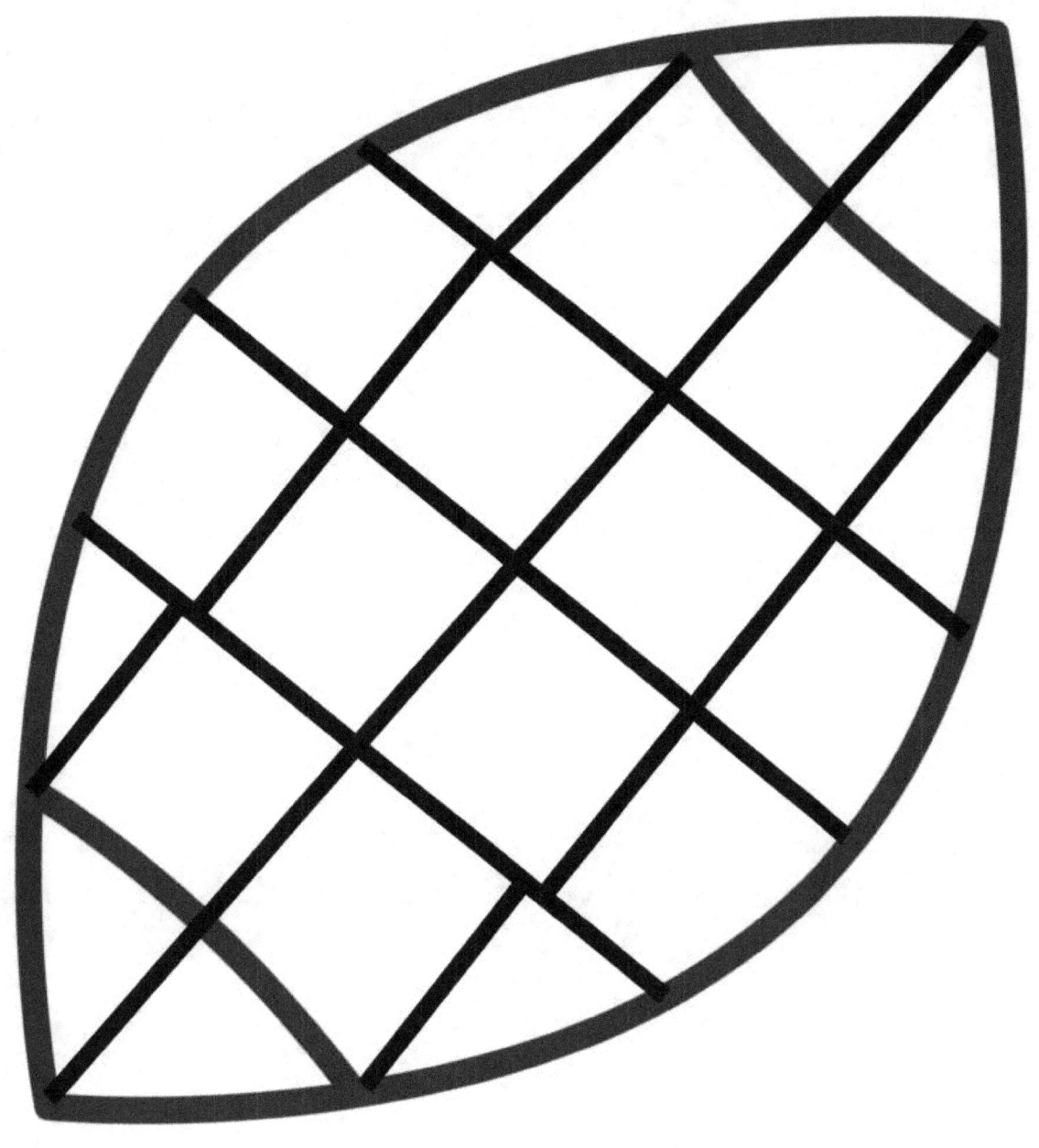

Use the puzzle mat to complete the puzzle.

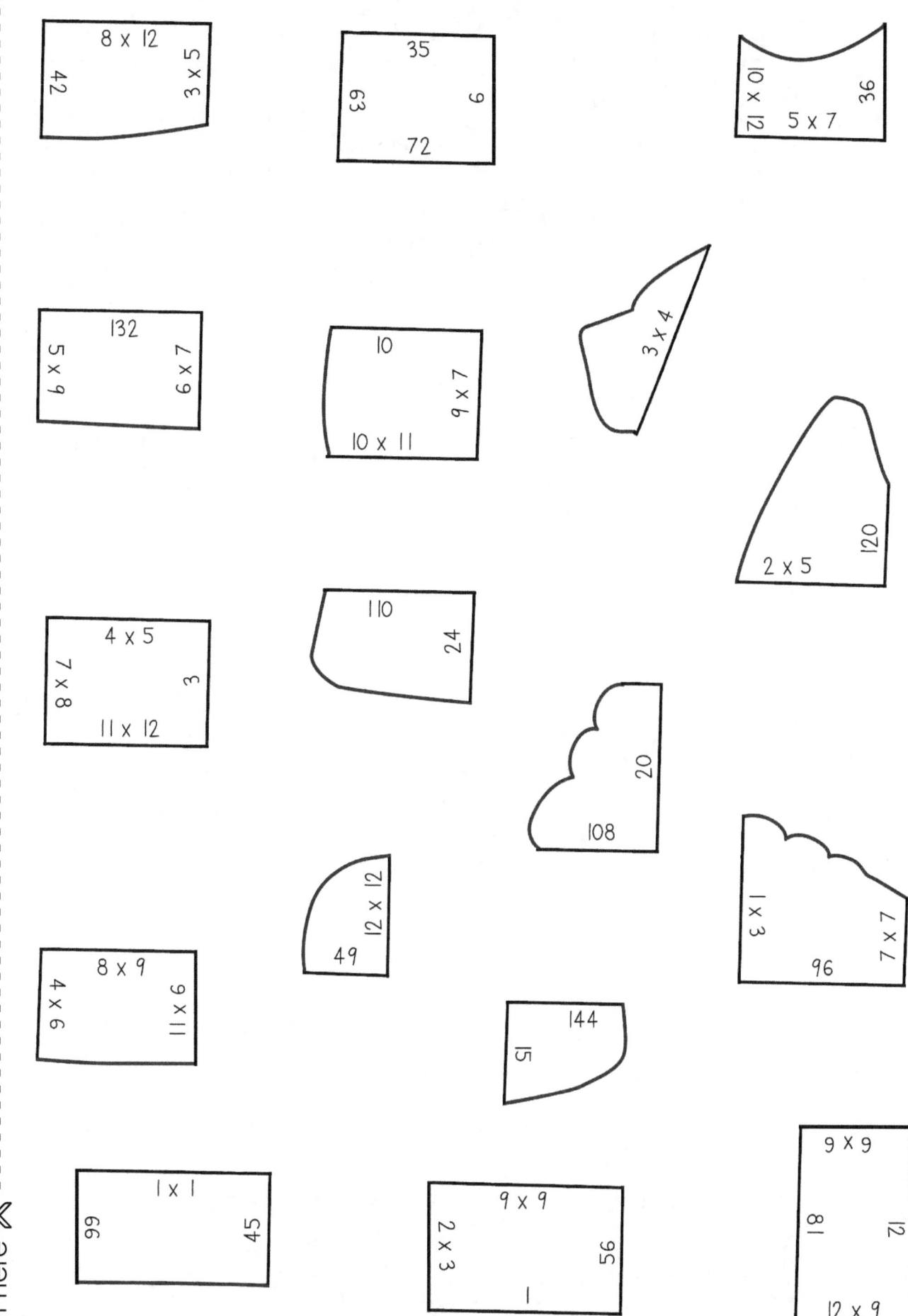

8 x 12
3 x 5
42

35
63
9
72

10 x 12
5 x 7
36

132
5 x 9
6 x 7

10
9 x 7
10 x 11

3 x 4

120
2 x 5

4 x 5
7 x 8
3
11 x 12

110
24

20
108

8 x 9
4 x 6
11 x 9

12 x 12
49

1 x 3
7 x 7
96

144
15

1 x 1
99
45

9 x 9
2 x 3
56
1

9 x 9
81
12
12 x 9

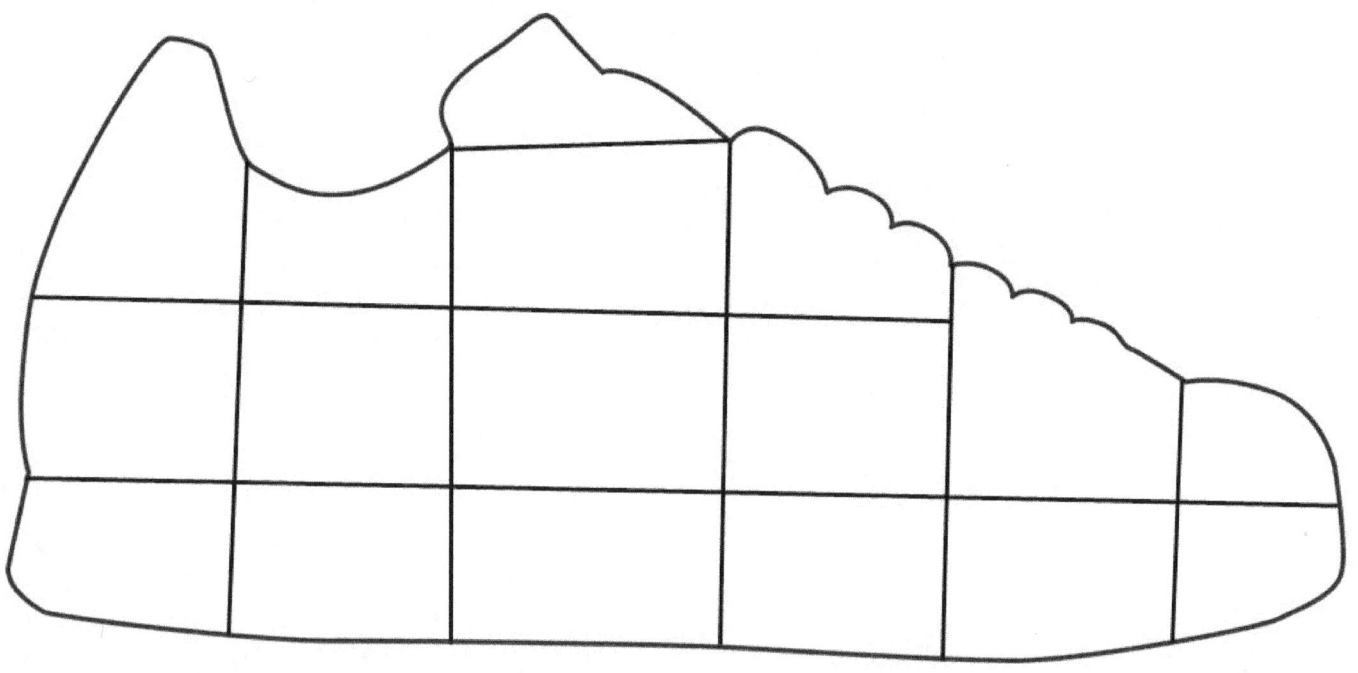

Use the puzzle mat to complete the puzzle.

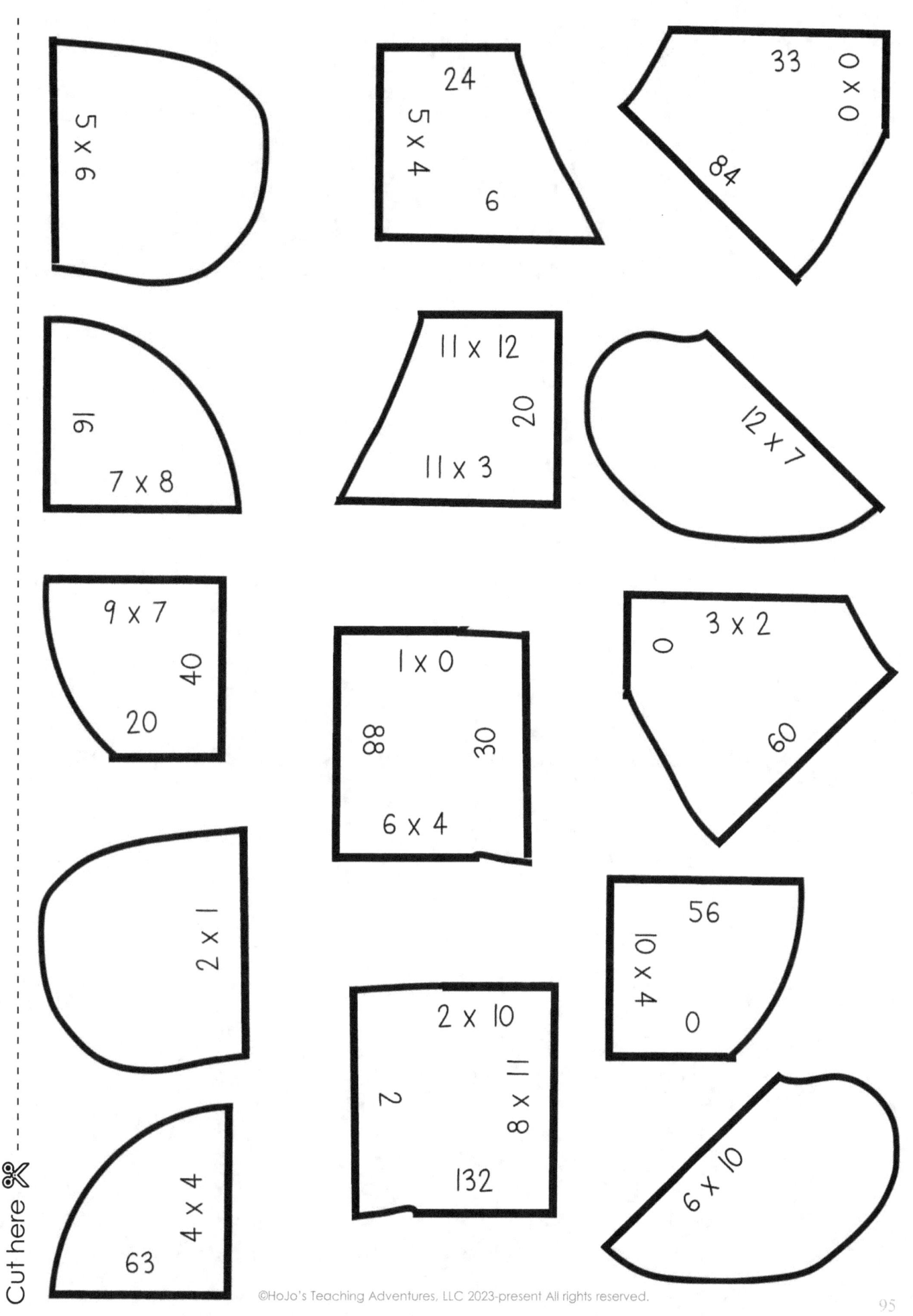

Use the puzzle mat to complete the puzzle.

4 x 10

7 x 6

27

40

80

9 x 8

35

6 x 9

54

3 x 9

11 x 9

8 x 10

64

18

2 x 9

5 x 7

42

99

132

72

8 x 8

12 x 11

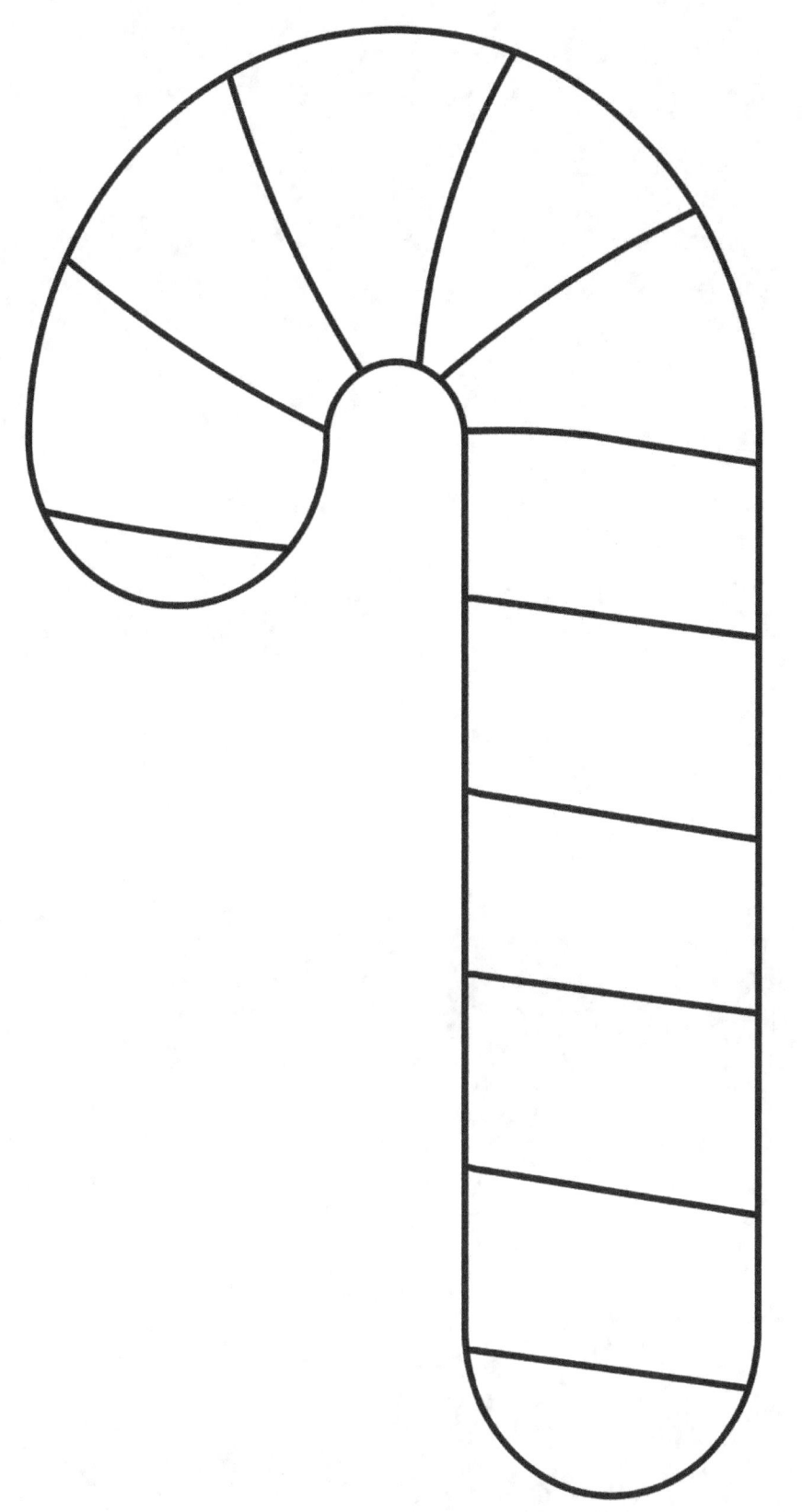

Use the puzzle mat to complete the puzzle.

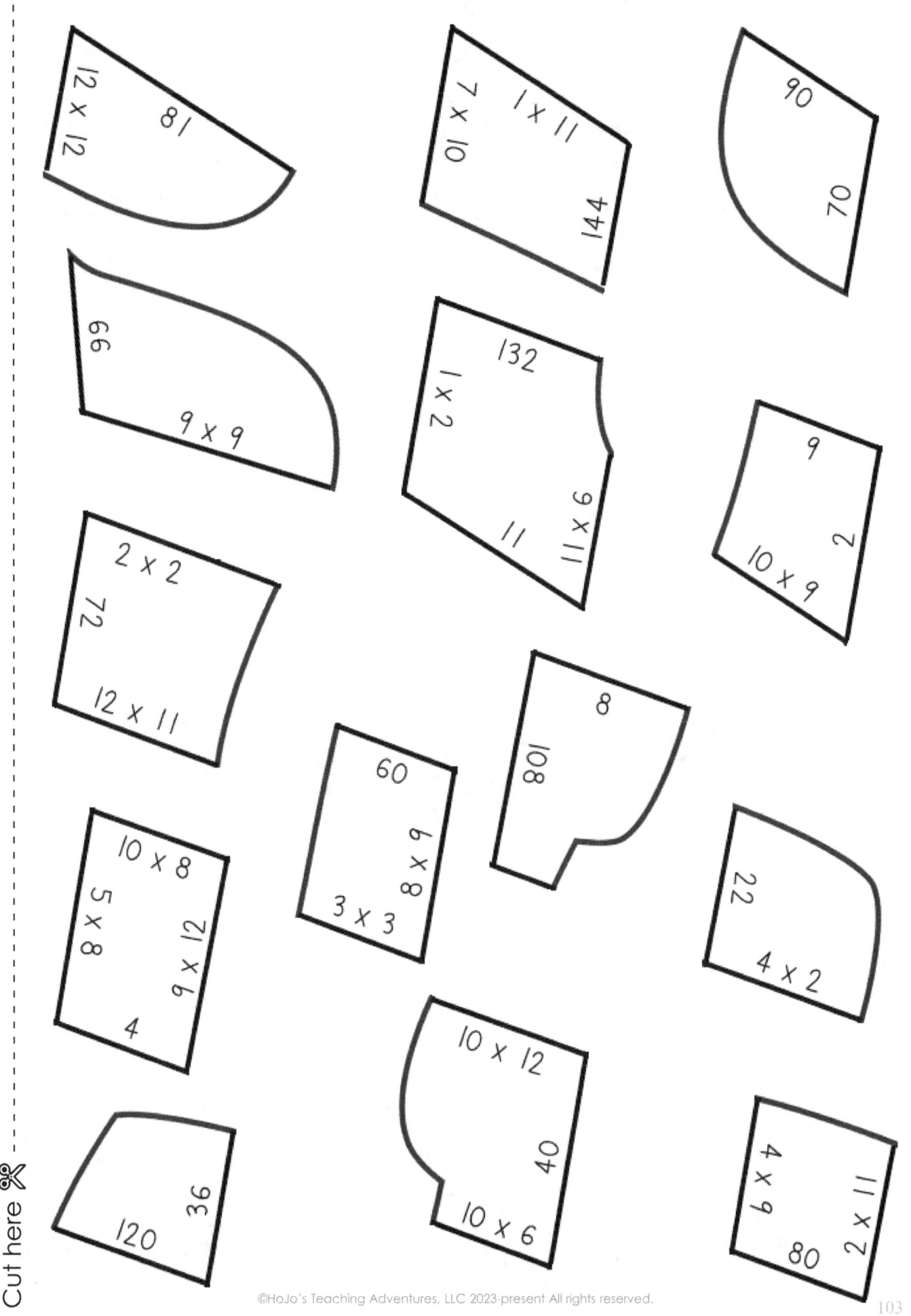

12 x 12
81

7 x 10
1 x 11
144

90
70

66
9 x 9

1 x 2
132
9 x 11
11

9
2
10 x 9

2 x 2
72
12 x 11

60
9 x 8
3 x 3

8
108

22
4 x 2

10 x 8
5 x 8
9 x 12
4

10 x 12
40
10 x 6

4 x 9
2 x 11
80

36
120

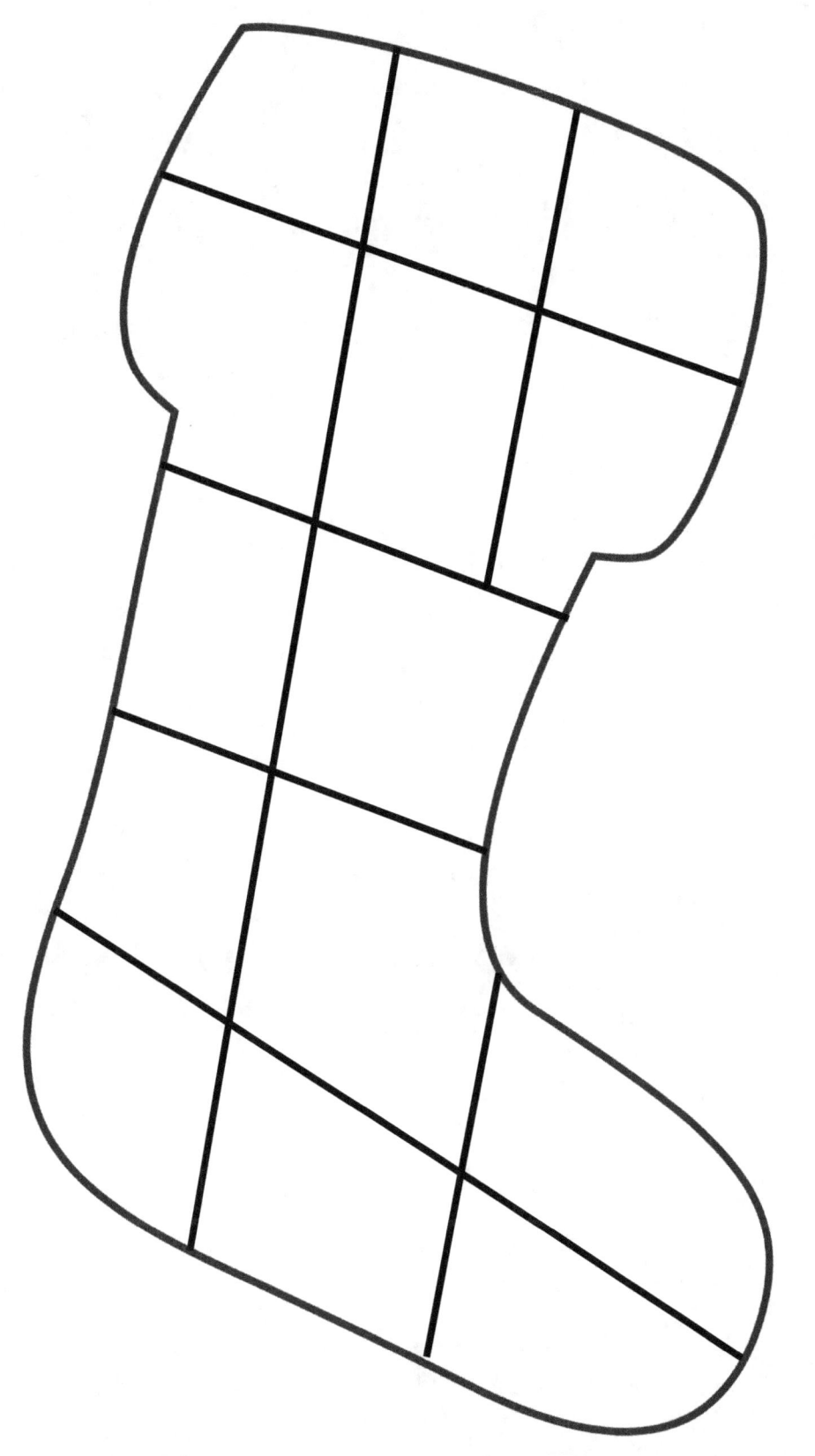

Use the puzzle mat to complete the puzzle.

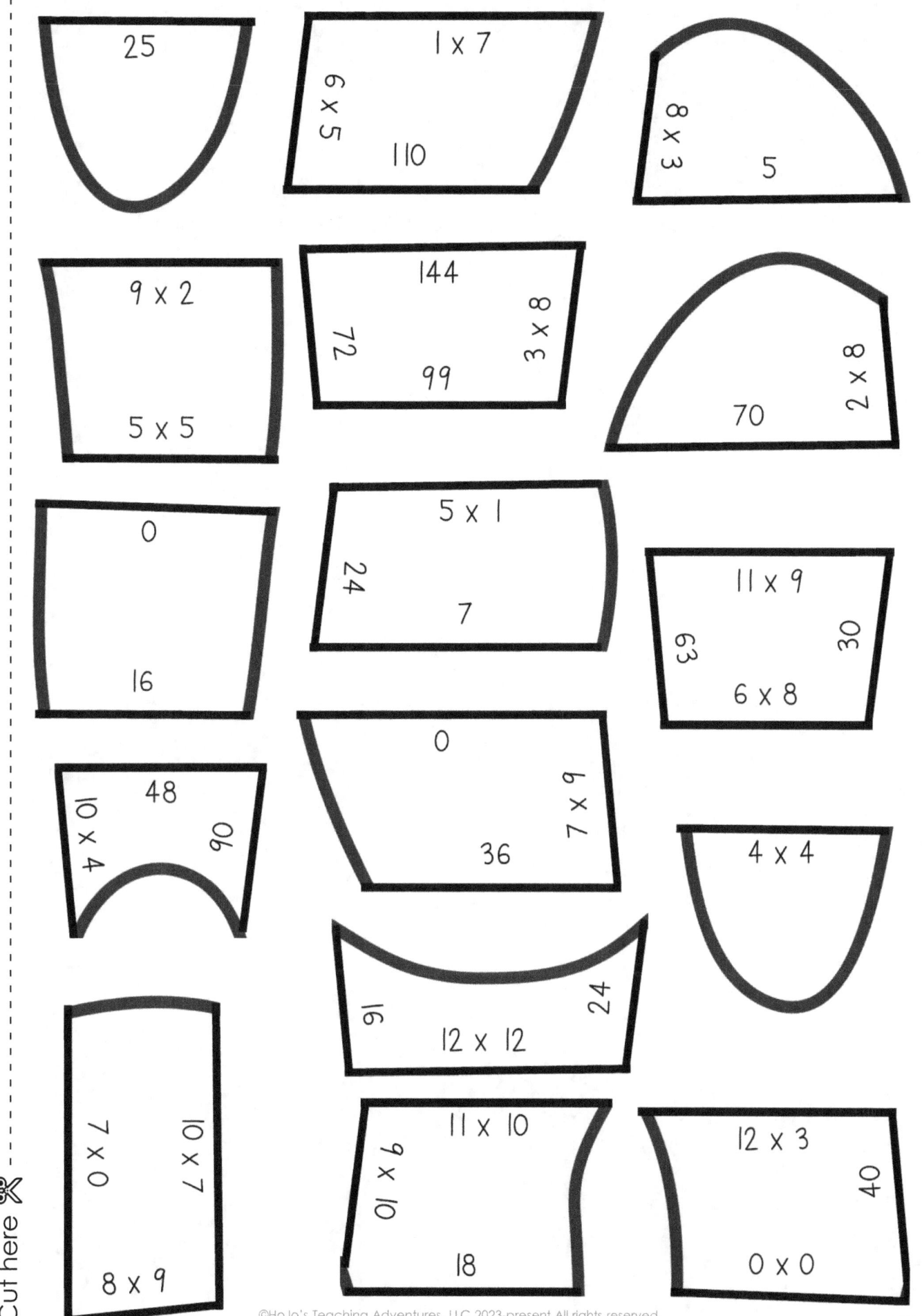

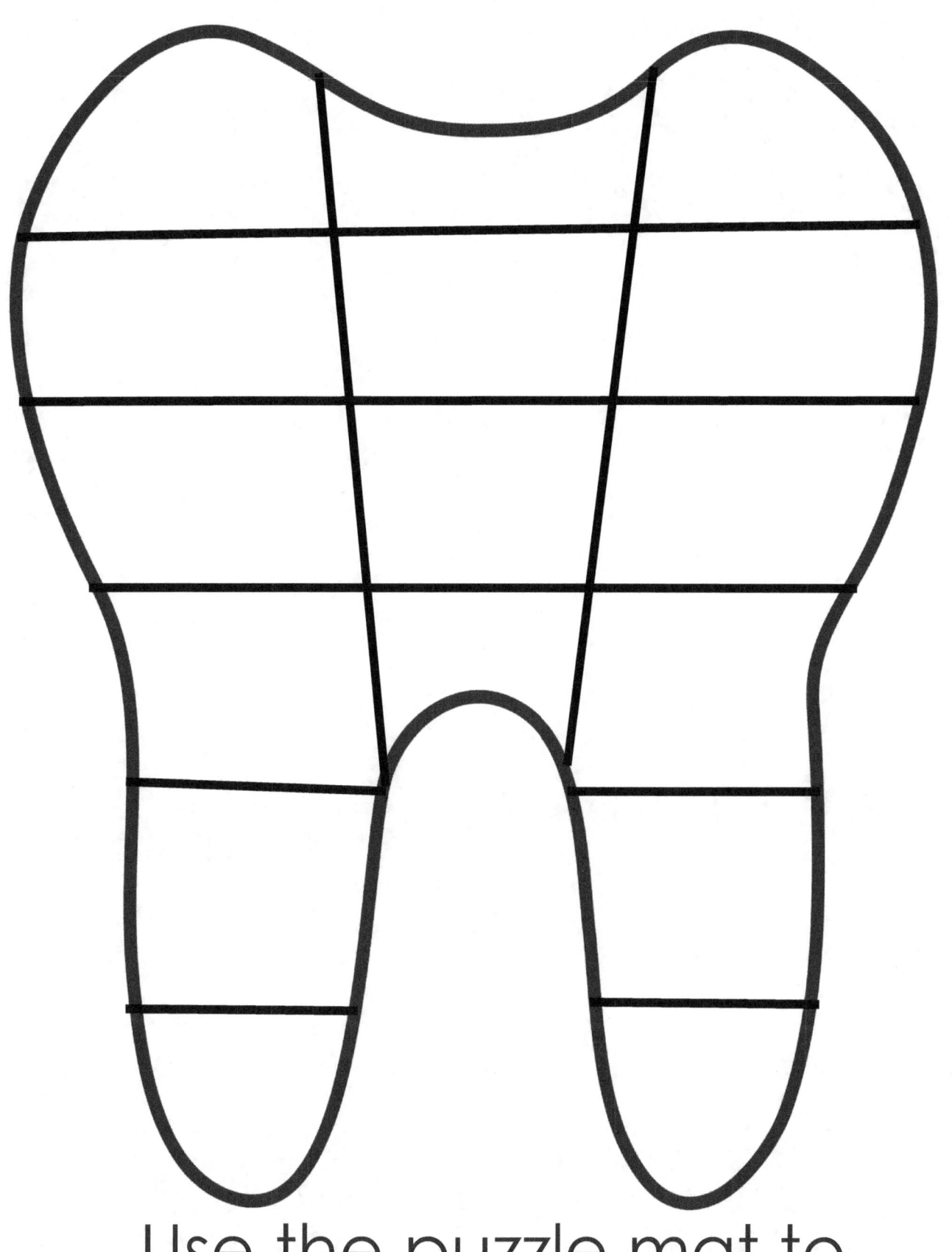

Use the puzzle mat to complete the puzzle.

8
36
6 x 3
4 x 10

8 x 7
9
30
8 x 3

4 x 7
56

81
5 x 6
35
1 x 8

10 x 2
21

18
4 x 3

2 x 2
80

20
3 x 2

28
9 x 9

40
10 x 8

24
7 x 3
9 x 4
4

7 x 5
12

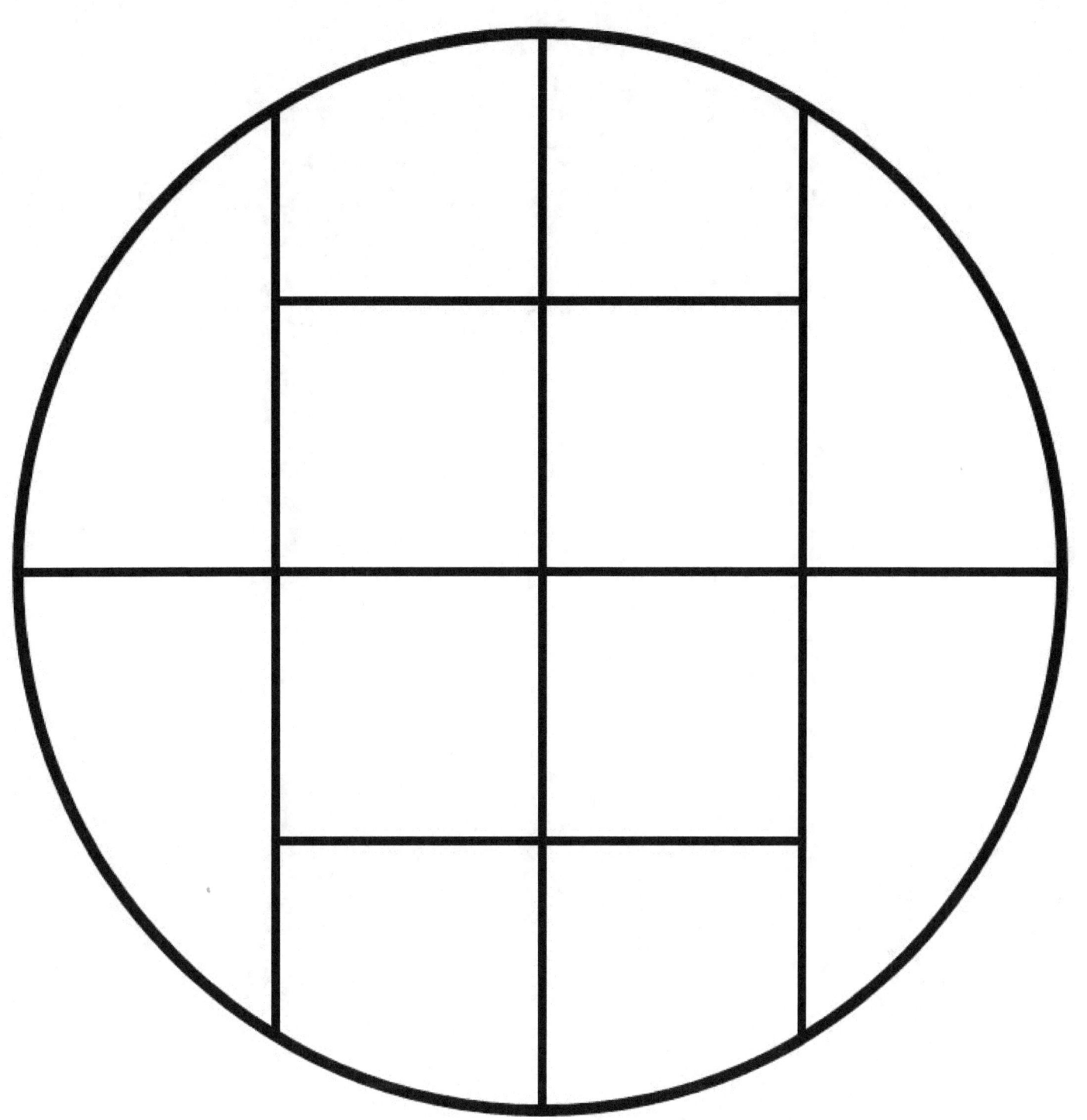

Use the puzzle mat to complete the puzzle.

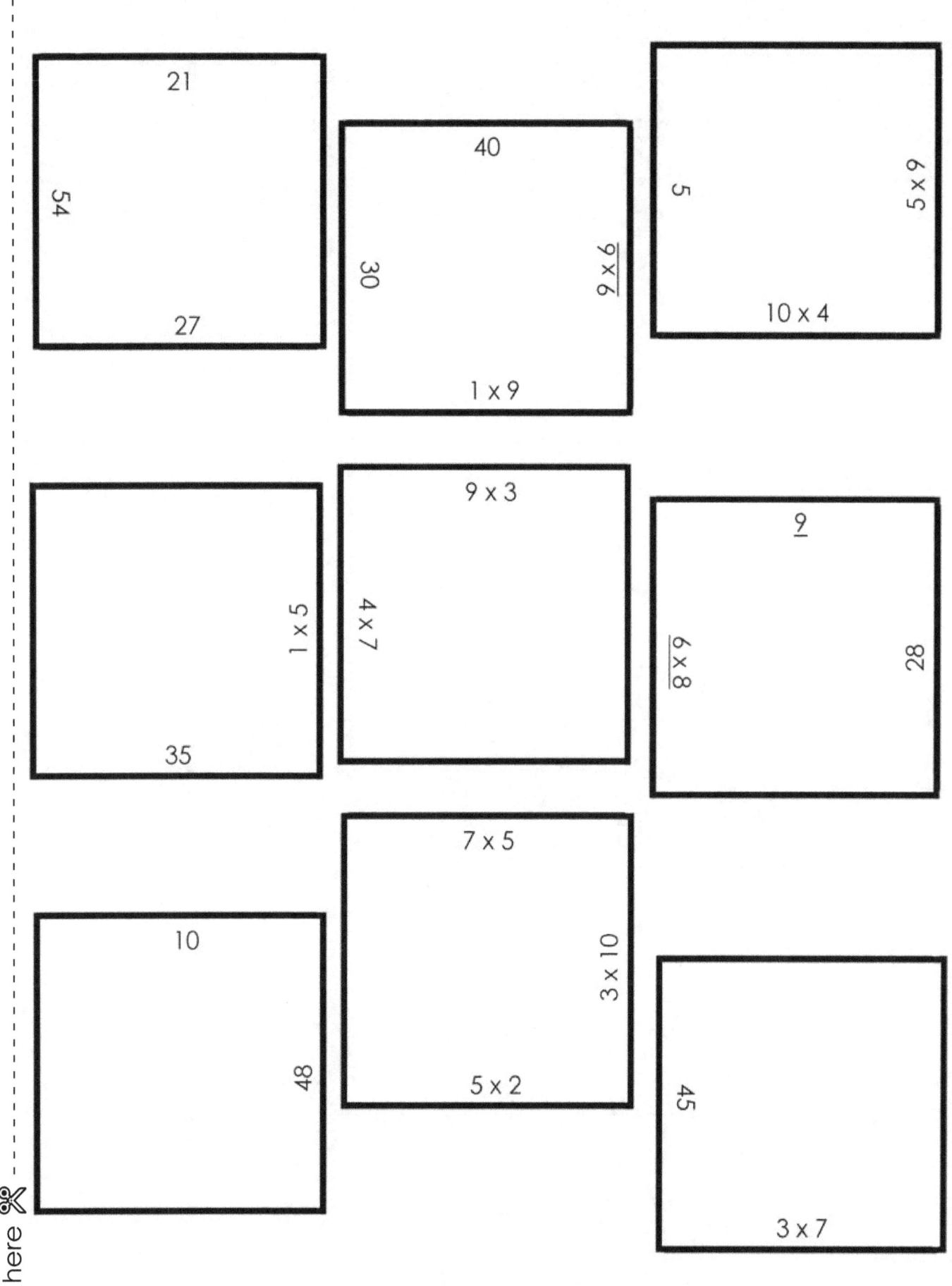

21

54

27

40

9 x 9

30

1 x 9

5 x 9

5

10 x 4

9 x 3

4 x 7

1 x 5

35

9

6 x 8

28

7 x 5

3 x 10

5 x 2

10

48

45

3 x 7

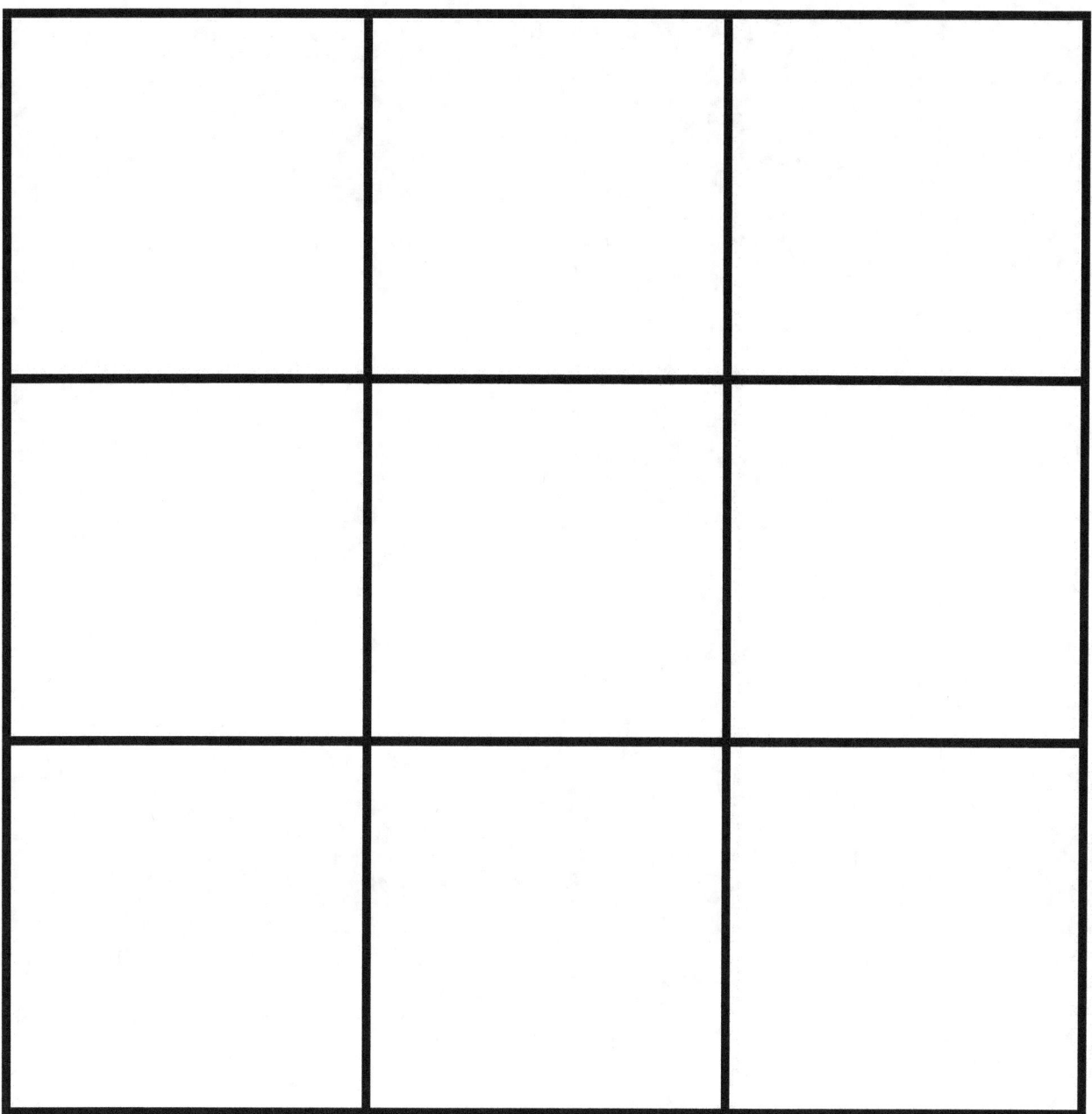

Use the puzzle mat to complete the puzzle.

Square (top left): 40, 18, 10 x 3

Square (top right): 7 x 1, 30, 2 x 8

Square (middle): 32, 6, 90, 7

Triangle: 9 x 10, 16

Triangle: 72, 3 x 6

Square: 7 x 9, 8 x 9, 1 x 6, 5 x 8

Triangle: 6 x 9, 2 x 2

Trapezoid: 70, 2 x 4, 4 x 8

Triangle: 36, 10 x 7

Trapezoid: 4, 8, 63

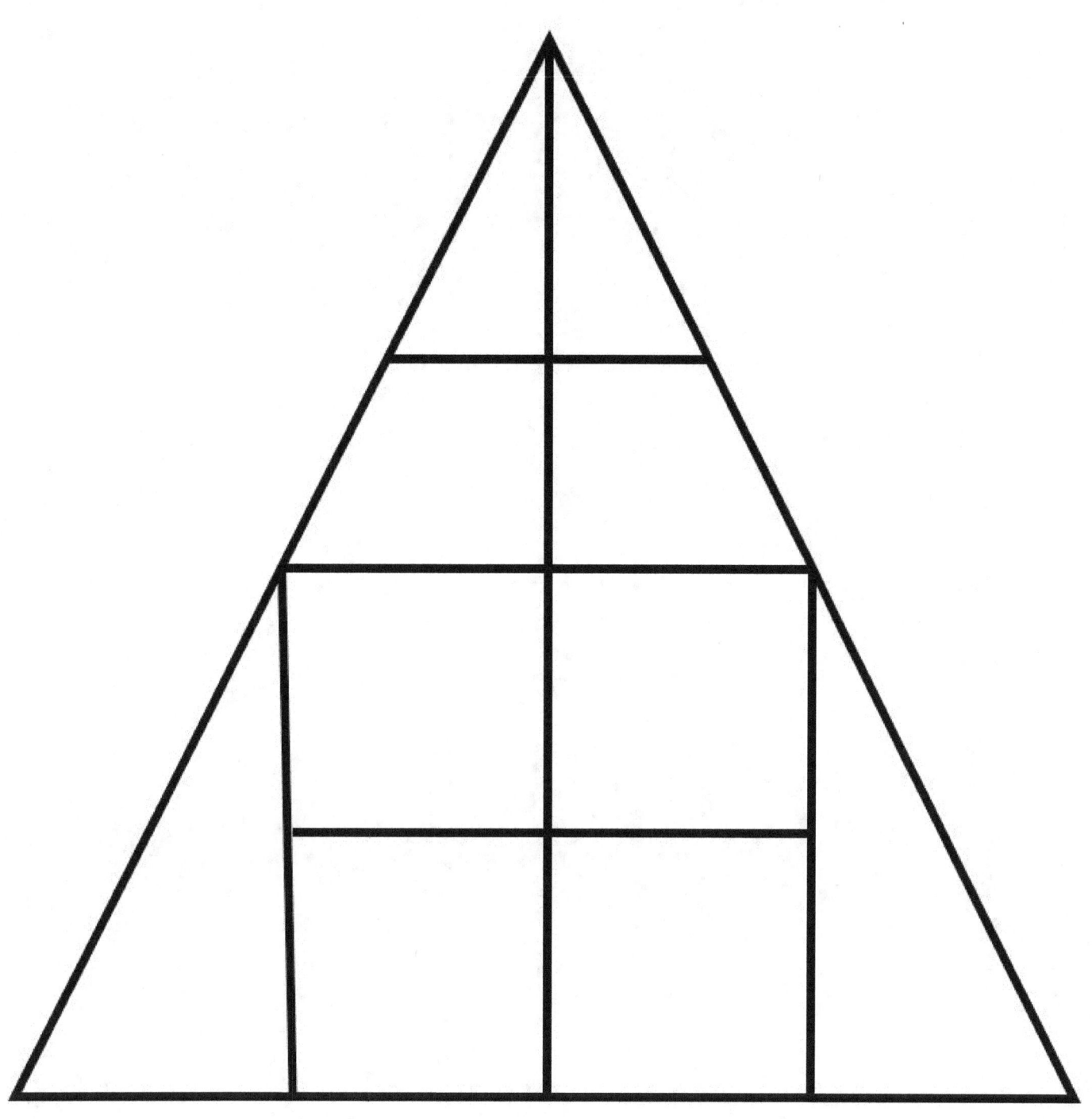

Use the puzzle mat to complete the puzzle.

Use the puzzle mat to complete the puzzle.

Multiplication Charts

1
1 x 1 = 1
1 x 2 = 2
1 x 3 = 3
1 x 4 = 4
1 x 5 = 5
1 x 6 = 6
1 x 7 = 7
1 x 8 = 8
1 x 9 = 9
1 x 10 = 10
1 x 11 = 11
1 x 12 = 12

2
2 x 1 = 2
2 x 2 = 4
2 x 3 = 6
2 x 4 = 8
2 x 5 = 10
2 x 6 = 12
2 x 7 = 14
2 x 8 = 16
2 x 9 = 18
2 x 10 = 20
2 x 11 = 22
2 x 12 = 24

3
3 x 1 = 3
3 x 2 = 6
3 x 3 = 9
3 x 4 = 12
3 x 5 = 15
3 x 6 = 18
3 x 7 = 21
3 x 8 = 24
3 x 9 = 27
3 x 10 = 30
3 x 11 = 33
3 x 12 = 36

4
4 x 1 = 4
4 x 2 = 8
4 x 3 = 12
4 x 4 = 16
4 x 5 = 20
4 x 6 = 24
4 x 7 = 28
4 x 8 = 32
4 x 9 = 36
4 x 10 = 40
4 x 11 = 44
4 x 12 = 48

5
5 x 1 = 5
5 x 2 = 10
5 x 3 = 15
5 x 4 = 20
5 x 5 = 25
5 x 6 = 30
5 x 7 = 35
5 x 8 = 40
5 x 9 = 45
5 x 10 = 50
5 x 11 = 55
5 x 12 = 60

6
6 x 1 = 6
6 x 2 = 12
6 x 3 = 18
6 x 4 = 24
6 x 5 = 30
6 x 6 = 36
6 x 7 = 42
6 x 8 = 48
6 x 9 = 54
6 x 10 = 60
6 x 11 = 66
6 x 12 = 72

7
7 x 1 = 7
7 x 2 = 14
7 x 3 = 21
7 x 4 = 28
7 x 5 = 35
7 x 6 = 42
7 x 7 = 49
7 x 8 = 56
7 x 9 = 63
7 x 10 = 70
7 x 11 = 77
7 x 12 = 84

8
8 x 1 = 8
8 x 2 = 16
8 x 3 = 24
8 x 4 = 32
8 x 5 = 40
8 x 6 = 48
8 x 7 = 56
8 x 8 = 64
8 x 9 = 72
8 x 10 = 80
8 x 11 = 88
8 x 12 = 96

9
9 x 1 = 9
9 x 2 = 18
9 x 3 = 27
9 x 4 = 36
9 x 5 = 45
9 x 6 = 54
9 x 7 = 63
9 x 8 = 72
9 x 9 = 81
9 x 10 = 90
9 x 11 = 99
9 x 12 = 108

10
10 x 1 = 10
10 x 2 = 20
10 x 3 = 30
10 x 4 = 40
10 x 5 = 50
10 x 6 = 60
10 x 7 = 70
10 x 8 = 80
10 x 9 = 90
10 x 10 = 100
10 x 11 = 110
10 x 12 = 120

11
11 x 1 = 11
11 x 2 = 22
11 x 3 = 33
11 x 4 = 44
11 x 5 = 55
11 x 6 = 66
11 x 7 = 77
11 x 8 = 88
11 x 9 = 99
11 x 10 = 110
11 x 11 = 121
11 x 12 = 132

12
12 x 1 = 12
12 x 2 = 24
12 x 3 = 36
12 x 4 = 48
12 x 5 = 60
12 x 6 = 72
12 x 7 = 84
12 x 8 = 96
12 x 9 = 108
12 x 10 = 120
12 x 11 = 132
12 x 12 = 144

11s and 12s Multiplication Charts

11	12
11 x 1 = 11	12 x 1 = 12
11 x 2 = 22	12 x 2 = 24
11 x 3 = 33	12 x 3 = 36
11 x 4 = 44	12 x 4 = 48
11 x 5 = 55	12 x 5 = 60
11 x 6 = 66	12 x 6 = 72
11 x 7 = 77	12 x 7 = 84
11 x 8 = 88	12 x 8 = 96
11 x 9 = 99	12 x 9 = 108
11 x 10 = 110	12 x 10 = 120
11 x 11 = 121	12 x 11 = 132
11 x 12 = 132	12 x 12 = 144

ANSWER KEY – Page 1

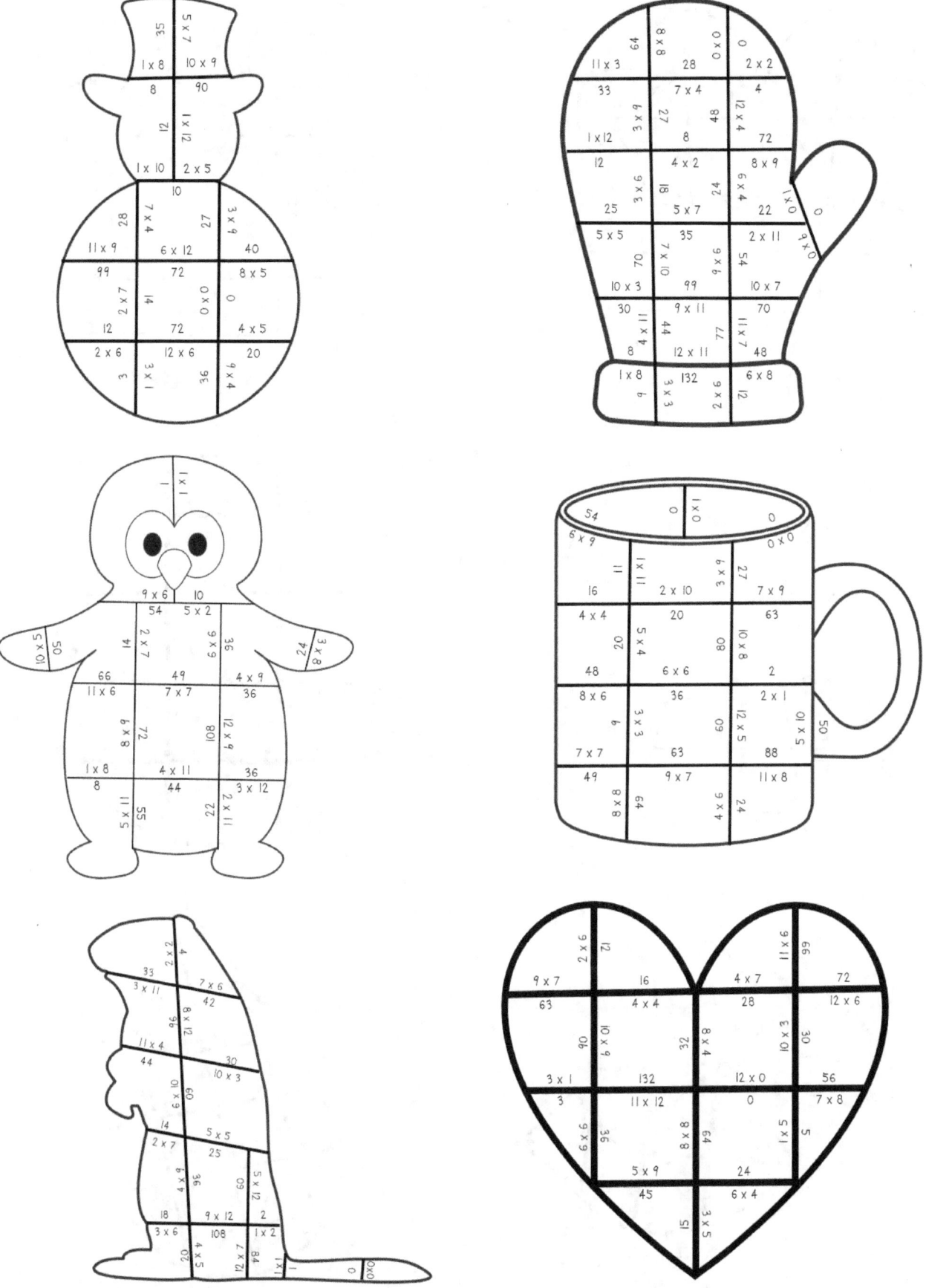

ANSWER KEY – Page 2

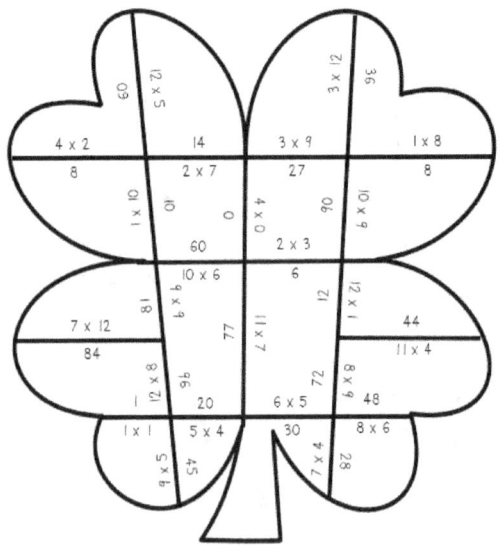

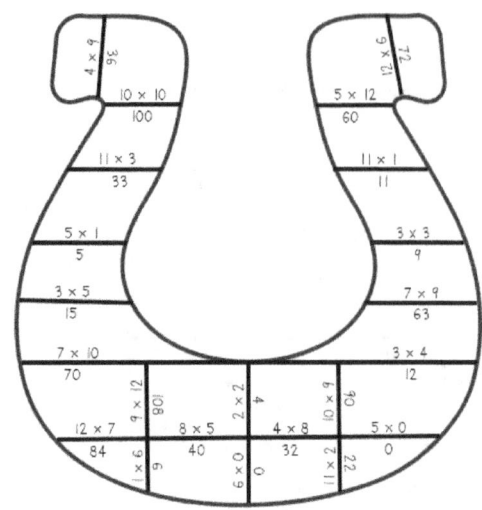

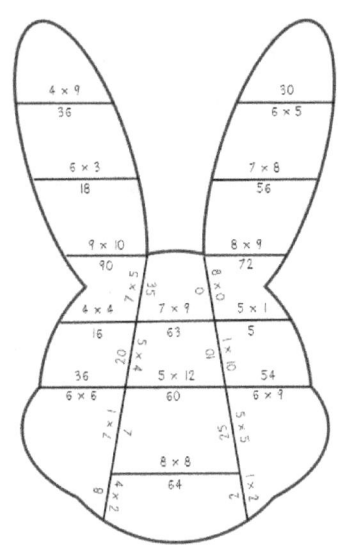

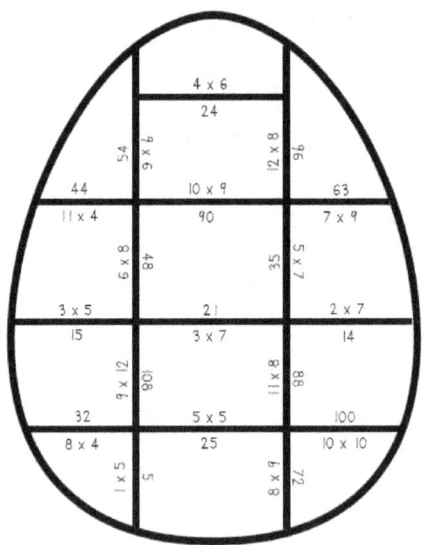

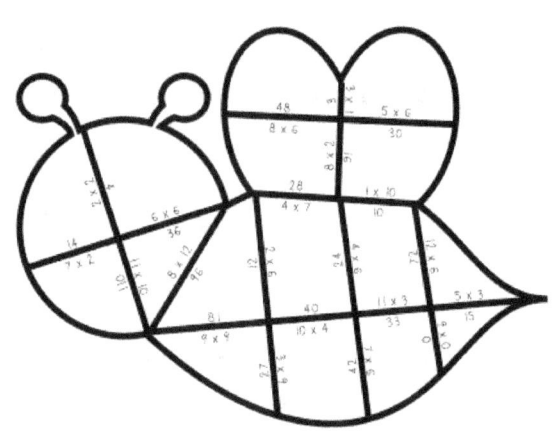

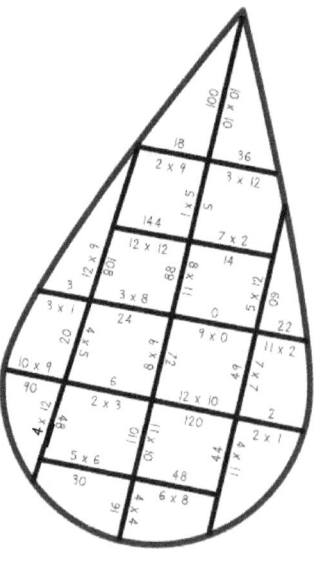

ANSWER KEY – Page 3

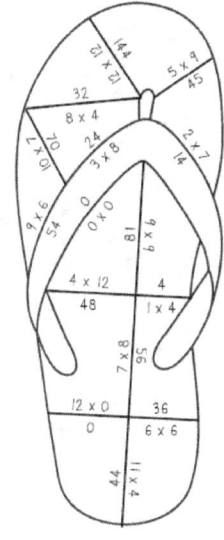

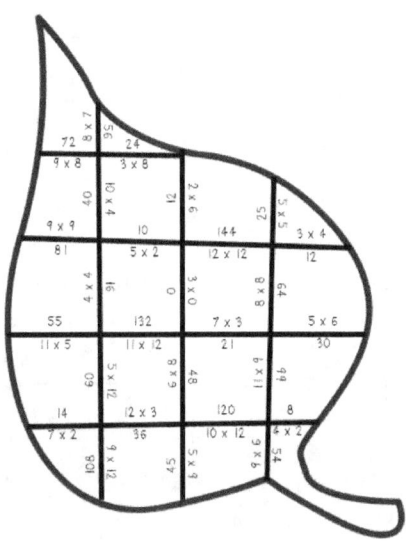

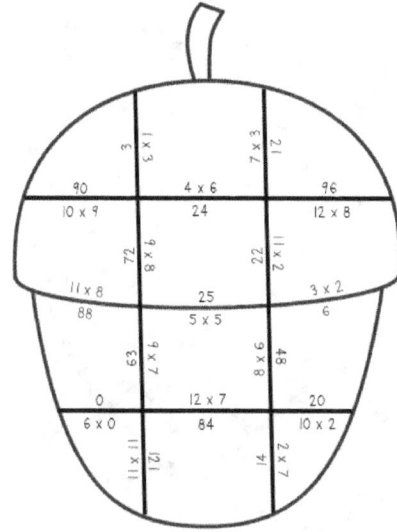

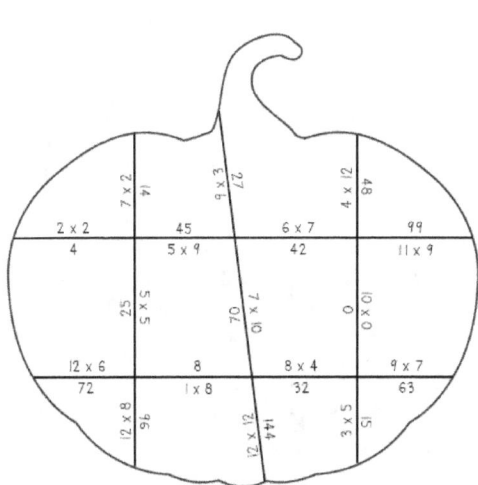

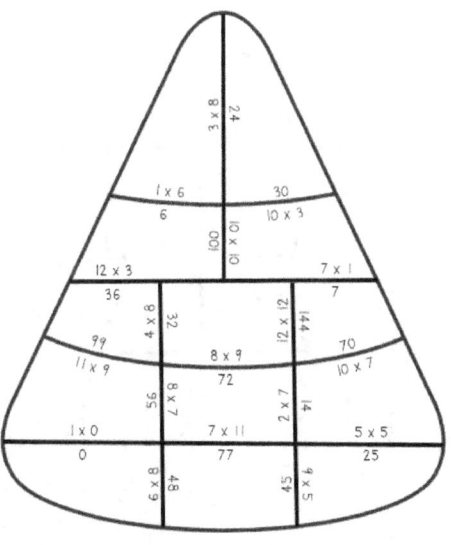

ANSWER KEY – Page 4

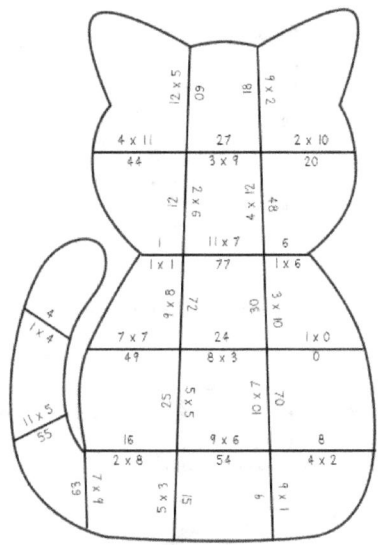

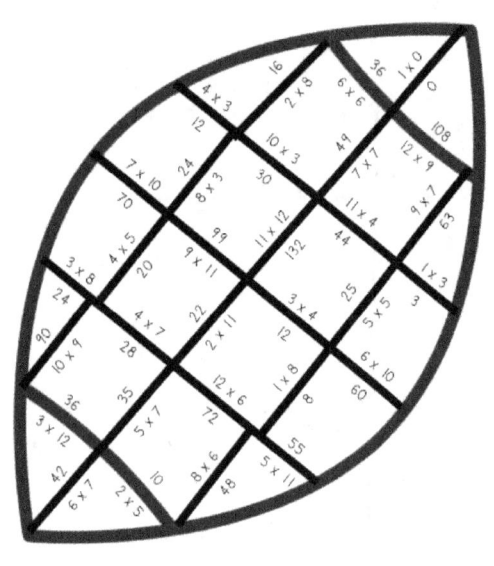

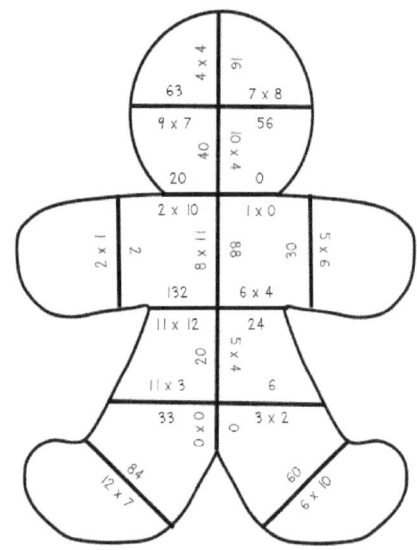

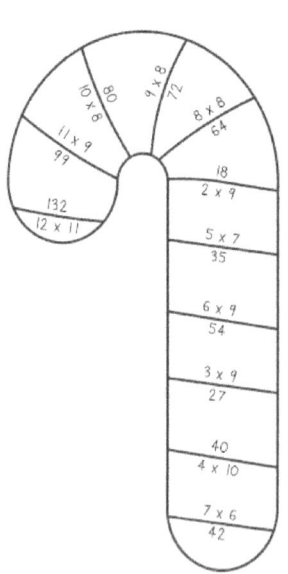

ANSWER KEY – Page 5

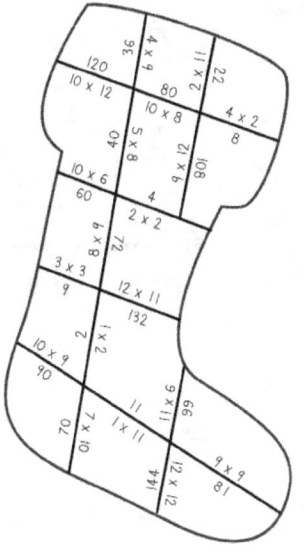

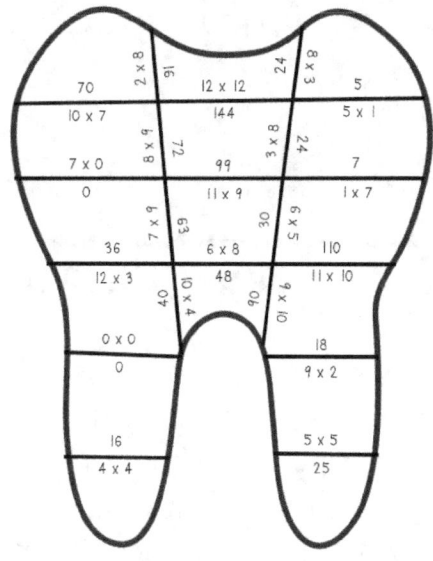

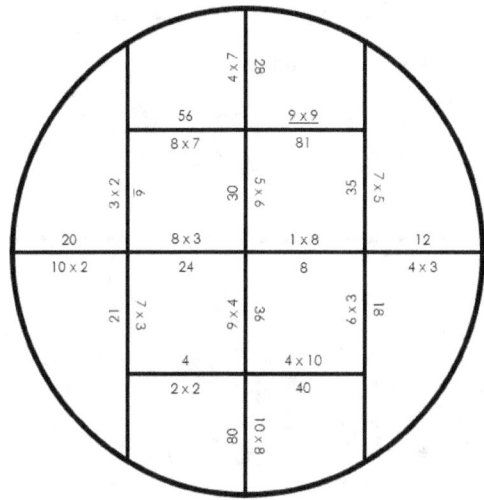

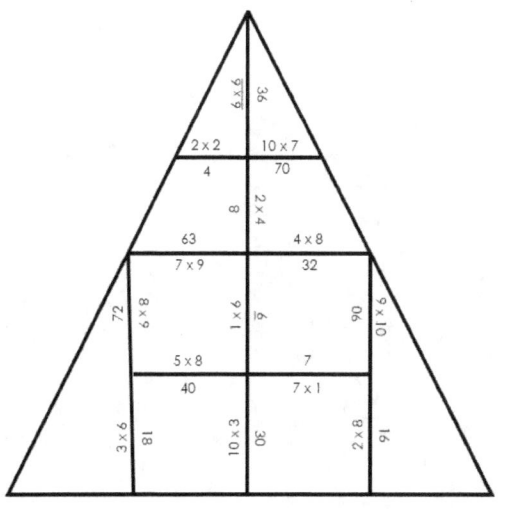

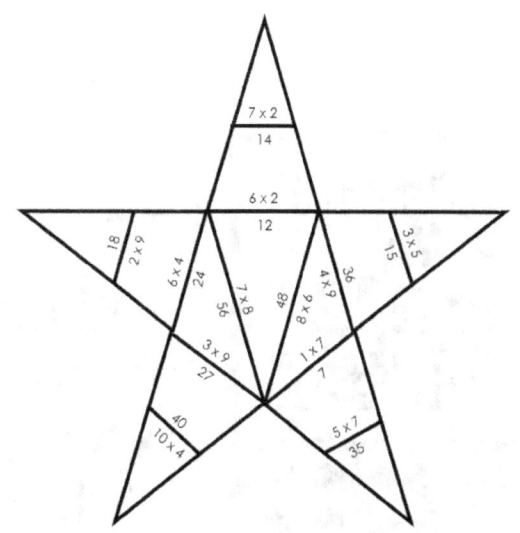

www.HoJosTeachingAdeventures.com

Want FREE math puzzles you can use today?

SCAN ME